"*Defending Beef* is a no-nonsense, scientific yet holistic look at the important role well-raised meat has in our food system and in ecosystem function. Nicolette Hahn Niman intelligently busts the common misperceptions about cattle and explains how, when managed properly, they can have a positive impact on the environment."

—**DIANA RODGERS**, registered dietitian, filmmaker and coauthor of *Sacred Cow: The Case for (Better) Meat*

"With all the rhetoric we are hearing today about how cows are destroying the planet, it is enlightening to hear the truth. Nicolette Hahn Niman delves deep into the science and sets the record straight: 'It's not the *cow*, it's the *how*'! Now, consumers can enjoy the health benefits of nutrient-dense beef while healing our ecosystems! A must-read for consumers, ranchers, and policymakers."

—**GABE BROWN**, regenerative rancher and author of *Dirt to Soil*

"The original edition of *Defending Beef* offered a compendium of everything a person should know about the role of beef cattle on the landscape and in our diets. This brand-new edition is more like a meta-analysis, chock-full of references, that dismantles almost every argument made against the ecological and nutritional importance of beef. While Nicolette Hahn Niman decries the industrial beef model, she makes a clear and compelling case why well-managed cattle grazing is a critical tool for capturing carbon and turning nonedible plant material into protein, as well as for supporting regenerative farming methods. This book should be on the shelf of anyone who cares about our climate and food system."

—**REBECCA THISTLETHWAITE**, coauthor of *The New Livestock Farmer*

"In this exhaustive and well-documented treatise, Nicolette Hahn Niman manages to be both informative and engaging from cover to cover. I especially appreciate the long myth-busting section that debunks many oft-cited anti-beef studies. This is the perfect book to have at your fingertips when you're in a dispute with someone who thinks meatless lab burgers are a great way to go."

—**RIDGE SHINN**, founding CEO, Big Picture Beef

"In this remarkable book, Nicolette Hahn Niman proves herself to be a true environmentalist—one who is willing to dig deeply, challenge orthodoxies, and get to the truth. You should read *Defending Beef* not only for the compelling case she makes for sustainable meat production, but also as an example of critical thinking at its finest."

—**BO BURLINGHAM**, editor-at-large of *Inc.* magazine and author of *Small Giants* and *Finish Big: How Great Entrepreneurs Exit Their Companies on Top*

"I have traveled to every state in the U.S. during both summer and winter and have seen the land in extensive rural areas. There are huge land areas in this country that cannot be used for crops. The only way to grow food on these lands is by grazing animals. Grazing done properly will improve the land. *Defending Beef* shows clearly that beef cattle are an important part of sustainable agriculture."

—**TEMPLE GRANDIN**, author of *Animals Make Us Human* and professor of animal science, Colorado State University

"Anyone hesitating to eat beef due to environmental or nutritional concerns needs to learn the other side of the story. *Defending Beef* is both scientifically accurate and highly readable. Kudos to Nicolette Hahn Niman for successfully engaging in one of the biggest environmental tensions of our day."

—**JOEL SALATIN**, farmer and author

"Creating healthful, delicious food in ecological balance is among humanity's greatest challenges. In this insightful book, Nicolette Hahn Niman shows why cattle on grass are an essential element. Every chef in America should read this book."

—**ALICE WATERS**, founder/owner, Chez Panisse, and author of *We Are What We Eat*

"Anyone who doubts that beef can be part of a sustainable food system and healthy diet should read this book. *Defending Beef* proves beyond a shadow of a doubt that we can feel good about eating beef that's raised the right way."

—**STEVE ELLS**, founder and CEO, Chipotle Mexican Grill

"Nicolette Hahn Niman just became beef's most articulate advocate. In *Defending Beef,* she pivots gracefully between the personal and the scientific, the impassioned and the evenhanded. It's a deeply compelling and delicious vision for the future of food."

—**DAN BARBER**, author of *The Third Plate*

"*Defending Beef* is a brave, clear-headed, and necessary addition to the discussion about sustainable food systems. Using hard data and solid scientific research, Nicolette Hahn Niman, a lawyer turned rancher, presents a convincing case that everything we thought we knew about the environmental and human health damage caused by beef is just plain wrong."

—**BARRY ESTABROOK**, author of *Tomatoland: How Modern Industrial Agriculture Destroyed Our Most Alluring Fruit*

"The prosecution will never rest after the case presented here by this unusually well-armed defense lawyer. Exactly how much and in what ways cattle benefit our world—whether or not we eat beef—have never been more thoroughly explained. Cattle are lucky to have such a remarkable rancher gal come to their aid on our behalf."

—**BETTY FUSSELL**, author of *Raising Steaks: The Life and Times of American Beef*

"Nicolette Hahn Niman's *Defending Beef* is as timely as it is necessary. With patience and passion she separates truth from fiction in the emotional debate about the role of beef in our lives and the effect of its production on our planet. Far from being a bogeyman of climate change and other environmental concerns, she argues, cattle, when properly managed, can play an important role in local food systems, land health, and carbon sequestration. The key is treating cattle as an ally, not an enemy, and exploring opportunities instead of simply pointing fingers. In this exploration, *Defending Beef* leads the way!"

— **COURTNEY WHITE**, founder, Quivira Coalition, and author of *Grass, Soil, Hope*

"In our collective confusion and desperation about the environment, many zero in on cattle as an unlikely culprit for everything from water pollution to climate change. In *Defending Beef,* author, rancher, and environmental lawyer Nicolette Hahn Niman takes a nuanced look at the impact of livestock on land, water, the atmosphere, and human health. With clarity and eloquence, she puts research in context and shows that the raising of cattle can be destructive or restorative, depending on how the animals are managed. Cattle—and common sense—have found their champion."

— **JUDITH D. SCHWARTZ**, author of *Cows Save the Planet*

"Issues related to the long-term health effects of red meat, saturated fat, sugar, and grains are complex and I see the jury as still out on many of them. While waiting for the science to be resolved, Hahn Niman's book is well worth reading for its forceful defense of the role of ruminant animals in sustainable food systems."

— **MARION NESTLE**, professor of nutrition, food studies, and public health at New York University and author of *What to Eat*

"I hope this book, which is more about the future of humanity, will be read by every citizen—not just those who feel the need to defend their meat-eating preferences. Biologist, environmental lawyer, and mother Nicolette Hahn Niman has provided a balanced report on the effects of cattle production on our environment, health, and climate change. Openly accepting the damage done by modern-day cattle production— on the land and in factory feedlots—she effectively argues that cattle themselves are not the problem; it is the way they are being managed that is endangering our health, environment, and economy. We can do something about that, and we must for the sake of our children and grandchildren. Key to our success will be an informed citizenry—for whom this book will be an invaluable tool."

— **ALLAN SAVORY**, founder and president, the Savory Institute

Defending Beef

— Revised and Expanded Edition —

The Ecological and Nutritional Case for Meat

NICOLETTE HAHN NIMAN

Chelsea Green Publishing
White River Junction, Vermont
London, UK

Project Manager: Alexander Bullett
Project Editor: Benjamin Watson
Copy Editor: Laura Jorstad
Proofreader: Diane Durrett
Indexer: Linda Hallinger
Designer: Melissa Jacobson
Page Layout: Abrah Griggs

Printed in Canada.
First printing July 2021.
10 9 8 7 6 5 4 3 2 1   21 22 23 24 25

Our Commitment to Green Publishing

Chelsea Green sees publishing as a tool for cultural change and ecological stewardship. We strive to align our book manufacturing practices with our editorial mission and to reduce the impact of our business enterprise in the environment. We print our books and catalogs on chlorine-free recycled paper, using vegetable-based inks whenever possible. This book may cost slightly more because it was printed on paper that contains recycled fiber, and we hope you'll agree that it's worth it. *Defending Beef*, Revised and Expanded Edition, was printed on paper supplied by Marquis that is made of recycled materials and other controlled sources.

Library of Congress Cataloging-in-Publication Data
Names: Niman, Nicolette Hahn, author.
Title: Defending beef: the ecological and nutritional case for meat / Nicolette Hahn Niman.
Description: Revised and expanded edition. | White River Junction, Vermont: Chelsea Green
   Publishing, 2021. | Includes bibliographical references and index.
Identifiers: LCCN 2021012272 (print) | LCCN 2021012273 (ebook) | ISBN 9781645020141
   (paperback) | ISBN 9781645020158 (ebook)
Subjects: LCSH: Beef cattle—United States. | Ranching—Environmental aspects—United
   States. | Diet—United States.
Classification: LCC SF207 .N563 2021 (print) | LCC SF207 (ebook) | DDC 636.2/13—dc23
LC record available at https://lccn.loc.gov/2021012272
LC ebook record available at https://lccn.loc.gov/2021012273

Chelsea Green Publishing
85 North Main Street, Suite 120
White River Junction, Vermont USA

Somerset House
London, UK

www.chelseagreen.com

MIX
Paper from
responsible sources
FSC® C103567

For Miles and Nicholas

May you always appreciate cattle
for the food
and the life
they've provided you.

CONTENTS

PREFACE   xi

Introduction   1

Cattle: Environment and Culture

1. The Climate Change Case Against Cattle:
   Sorting Fact from Fiction   7
2. All Food Is Grass   59
3. Water   77
4. Biodiversity   100
5. Overgrazing   108
6. People   115

Beef: Food and Health

7. Health Claims Against Beef:
   The Rest of the Story   129
8. Beef Is Good Food   189

Critique and Final Analysis

A Critique: What's the Matter with Beef?   217
Final Analysis: Why Eat Animals?   224

ACKNOWLEDGMENTS   247
NOTES   248
INDEX   279

PREFACE

IN 1970, AMERICA'S GRASSROOTS ENVIRONMENTAL MOVEMENT was burgeoning as 20 million people poured into the streets on April 22 to mark the first Earth Day. Cattle raising was dragged into the public square as a villain along with our nation's worst industrial polluters. Beef was increasingly regarded as an ecosystem destroyer and a primary cause of starvation around the globe. It was becoming part of the zeitgeist to believe that no genuine environmentalist or humanitarian would eat beef (at least not in a well-lit public place). Kicked off by *Diet for a Small Planet*, three decades of influential books like *Diet for a New America* and *Beyond Beef* then succeeded in making it nearly incontrovertible environmental orthodoxy that beef is public enemy number one.

As a freshman biology major in the mid-1980s, I drank the Kool-Aid. I quit eating meat and enthusiastically embraced the attitude that no beef was good beef. Then I promptly filed the matter away; no more consideration of the topic seemed necessary.

That logic began fracturing in 2000, soon after I was hired as an environmental lawyer by Robert F. Kennedy Jr. He charged me with starting a national campaign to address meat industry pollution. At first, my assignment seemed to neatly reinforce my long-held negative views about meat and how it was produced. But the more farms I visited, the more studies I read, the more experts I interviewed, the less satisfied I became with my understanding of meat's connections to the environment. I began to recognize my views as simplistic—black-and-white and formed mostly from the bullet points in vegetarian and environmental pamphlets I'd encountered in college.

Fortunately, by the time a handsome cattle rancher proposed marriage to me two years later, my understanding of the role farm animals can and do play in food systems and natural environments was far more nuanced. And I had the good sense to accept him.

Working in the meadows and valleys of our ranch alongside my husband for the past seventeen years has given me an entirely new understanding of how ecosystems work. I've lived among not only cattle but also domesticated and wild turkeys, deer, coyotes, beetles, newts, bobcats, butterflies, ravens, hawks, egrets, gophers, and countless other animals, plants, and fungi. I've learned how humans can interact with ecosystems as landscapes that produce food while at the same time supporting—even *enhancing*—wild-living populations (including the unseen microscopic ones!) that belong here and form the foundation of all living systems.

Decades after that first Earth Day, a singularly negative view of ranching and beef persists among many environmentalists and those who oppose raising animals for food. It has leached into mainstream conversations and commentaries as global warming concerns have infused the issue with new life. As a lifelong environmentalist who unwaveringly followed a vegetarian diet for over three decades, I am intimately familiar with the criticisms. Rarely, though, have I encountered credible responses, least of all from the beef industry itself. Yet now, as a mother and a person more dedicated than ever to restoring our planet, while raising cattle myself, I feel compelled to respond, honestly and passionately. This is my answer. This is my defense of beef.

Introduction

WE'VE ALL HEARD THE NARRATIVE SO OFTEN—THE ONE ABOUT how red meat, beef in particular, is killing us—that many of us have come to accept it as incontrovertible truth. It's so common it's common knowledge. The story goes something like this:

Americans once raised cattle, pigs, and sheep on small, mixed farms scattered around the country, sprinkled with handfuls of livestock. Animal numbers were low and, correspondingly, Americans ate dainty portions of animal fat and red meat. We were thin. Hypertension, stroke, and heart disease rates were low. Environmental damage from farming was minimal. Over the course of the 20th century, however, everything changed for the worse. Livestock herds ballooned. Cattle overgrazed. Red meat and animal fat became abundant, cheap, and ubiquitous. Americans gorged themselves on hamburgers, butter, and ice cream. The result: soil erosion, water and air pollution, and skyrocketing rates of obesity and chronic diet-related diseases.

There's just one problem with this narrative: It's fiction.

Yes, parts are correct. But facts that rarely make it into mainstream discussions and media coverage diametrically oppose the narrative's key elements. As this book will make clear, aspects of the United States' environmental condition have, indeed, worsened, and chronic diet-related diseases have become more widespread and severe. But these problems cannot reasonably be connected with bovines, butter, or beef. Why?

Because there are about the same number of cattle on the land today as there were a century ago. And while Americans are taking in more calories overall, they are eating less red meat in general, and less beef in particular, than at any time in recent history. We are also consuming less butter, far less whole milk, and much less saturated animal fat. No swelling bovine herds. No ever-heftier helpings of red meat or beef fat. From these facts alone, the simplistic narrative collapses.

If you are skeptical, I won't blame you. What I've just said likely runs counter to what you've heard from various sources for many years. But I come armed with data, and plenty of it, all from official government sources. While my overall premises—that cattle are good for the environment and that beef, butter, and cheese are healthy foods—are, admittedly, controversial in this day and age, the basic agricultural and demographic facts are not in dispute.

Here is the most pertinent point to keep in mind. In the second half of this book, I will detail how American diets have changed. I will show we eat less beef and less animal fat now than we did 100 years ago, while our consumption rates for sugar, grain, and industrial vegetable oils have skyrocketed. I will present facts strongly supporting the conclusion that our sugar, flour, and vegetable oil consumption rather than red meat and animal fats are to blame for the sharp rise in obesity and chronic diseases.

The popular story line is also far off-base concerning the numbers of animals on the land. In reality, Americans' eating habits have shifted away from beef toward poultry and fish. Decreasing per-capita consumption of beef has been accompanied by a decline in per capita cattle in US inventory. It's true the total quantity of red meat and dairy produced has increased as our population has grown, and some is exported. But the amount of beef and dairy the United States exports is actually quite small. Only about 7 percent and 2 percent, respectively, go to foreign markets. So cattle raised for exported meat and milk products barely affect the math.

Greater output in the beef and dairy sectors has actually not resulted from swelling herd sizes. On the meat side, this is because animals are now slaughtered at much younger ages. At the dawn of the 20th century, a typical beef steer was sent to slaughter at four or five years of age.[1] Today, to lower costs, and enabled by grain feeding and growth hormones, that steer is killed younger than two years old, typically around 14 months.[2] Dairy cows, too, go to slaughter at much younger ages (often just three years old). This also affects beef supply because, now, as always, a large portion of US beef comes from dairy cattle.

The rise in milk production, however, is owing to another issue. As I detailed in my book *Righteous Porkchop*, selective breeding of dairy cows for greater milk output (read: large bodies and huge udders) has vastly increased per-animal production.[3] At the beginning of the 20th century, average US annual per-cow milk output was 2,902 pounds (348 gallons). Today it is 22,774 pounds (2,734 gallons) per year. This is often touted as a major victory for humanity. But the scale of the increase (more than sevenfold) suggests selective breeding has been pushed to an extreme. (Indeed, many of today's mature dairy cows even have trouble walking, something I have personally witnessed, with heavy heart, numerous times.) The net effect of this change has been a substantial shrinkage of the US dairy herd over the past century.

These factors, combined with the near disappearance from American agriculture of oxen, mules, donkeys, and horses as animal draft power, means there are fewer larger farm animals in the United States now than there were a century ago.

For those of you who may still find this hard to believe, here are the specific numbers. Since 1900, beef cattle numbers rose, but less than people tend to assume, going from 67 million to 94 million. Pig numbers have also gone up, but not much: In 1920, there were 60 million pigs; in 2018, 74 million. On the other hand, sheep numbers plummeted, going from a high of 46 million in 1940 to 5 million today. Likewise, the dairy cow herd shrank dramatically, from 32 million down to 9 million, over the past century. And draft animals have gone from 22 million in 1900 to just 3 million in 2002. All told, that means that while early-20th-century farms and ranches had roughly 99 million head of cattle and 227 million larger animals (including cattle), today they have 103 million cattle and 185 million larger animals. That's a modest 4 percent increase in cattle numbers and an overall 19 percent *reduction* in larger farm animals.

From an environmental standpoint, two issues are most relevant: How many animals are in inventory, and, more important, by what methods are they being raised? These factors will largely determine ecological footprint—harmful or helpful. As we've just seen, there are only slightly more cattle today than for much of the past century. At the same time, cattle are being raised with more care: There is a burgeoning movement within agriculture to thoughtfully manage grazing. This is increasingly transforming animal impact into a cornerstone of regenerative agriculture.

From a diet and human health perspective, the central questions are: What's being consumed, and in what form? Are we eating real, whole, unprocessed vegetables, fruits, nuts, grains, eggs, fish, and meats or, in contrast, "edible food-like substances," as Michael Pollan calls them in his book *Food Rules*? Americans today consume less beef and butter but far more processed foods. Fast food, packaged snacks, and sweet drinks for starters. We all know a bowl of potato chips is junk food whereas a baked potato is satiating and nourishing. Similarly, a mass-produced hot dog is wholly different from a steak. How much processed food one eats is proving to be the most important variable for health. Whether a food is healthful or harmful is also affected by how it was raised.

It's essential to acknowledge these facts when we talk about meat. If we recognize that total cattle numbers have been fairly stable and beef consumption is down, it immediately casts serious doubt on the all-too-common narrative that blames cattle and beef for our current environmental and public health crises. I would not expect these facts, on their own, to dissolve the concerns of beef's critics. But I hope clarifying these questions at the outset will allow readers to consider this book with an open mind.

I strongly dispute the charge that cattle and beef are responsible for the globe's environmental and human health problems. But aspects of the popular narrative I mentioned at the outset are correct, and they need immediate and sustained attention. Serious environmental degradation the world over has been caused by agriculture, including the cattle and beef sectors. Like much of American agriculture, the beef industry has become too dependent on manmade inputs like insecticides, fertilizers, hormones, and other pharmaceuticals. Until just the past few years, there has also been widespread failure to grasp soil biology as the essential foundation of truly regenerative farming. Many involved in mainstream farming, including those raising cattle, have failed to prioritize generating nutrient-rich, health-sustaining human foods. Simply put, industrialization has radically altered the way humans farm and eat, much of it for the worse. These issues will be explored in this book.

Current US food and farm policies subsidize output. Much of what is incentivized are destructive practices. Only a tiny portion of subsidies encourage ecological farming methods. Those same policies foster excess production of the unhealthy foods we are already overconsuming and that are making people sick. All of which contributes to underemployment, cheap food, overeating, waste, and environmental damage.

This book is at once a defense of cattle and beef, and an indictment of aspects of modern agriculture and diets. Change in both arenas is urgently needed. The United States is the globe's top beef producer.[4] We can, and *should*, lead the world in forging ways of raising cattle that reverse environmental degradation and produce healthy, nutrient-rich foods.

For me, it comes down to this: Cattle are central to the human story. We have lived alongside them for tens of thousands of years. Our close connection with cattle has boosted our bodies' immunity, enabled our migrations, and provided us with intensely nourishing, delicious foods. Done with care, cattle husbandry enriches our human experience and enhances the natural world. We must move well beyond simplistic solutions like banishing cattle from our landscapes and beef and butter from our plates. Instead, it's time to focus on improving how we raise cattle and turn them into food. Only then can we tap into the full ecological and nourishment potential these remarkable creatures provide. As my friend Russ Conser has said, "It's not the *cow*, it's the *how*."

Whether you are among the critics or the defenders of beef, if you come along on this journey, you will find things to agree and disagree with along the way. Whatever your perspective at the start, I hope by the end you will see things in a new and different light.

CHAPTER 1

The Climate Change Case Against Cattle

Sorting Fact from Fiction

FOR DECADES, THE PRIMARY ENVIRONMENTAL OBJECTION TO BEEF was "overgrazing." Cattle, it was said, had devastated vast stretches of the American West and much of the globe with their munching mouths and trampling hooves. Damaged waterways, eroded and denuded landscapes, reduced wildlife populations, and, most worryingly, deserts everywhere spreading like wildfires were said to be the results. Deforestation, like that occurring in the Amazonian rain forest, was often tossed into the mix.

Two baseline notions undergirded the charges against cattle—things so obvious they needed no proof. First, trampling and grazing inherently cause environmental injury: The more grazing and trampling, the more damage. Second, the more cattle in an area, the graver the ecological harm. A bit later, I will show these beliefs are proving incorrect. For the moment, I just want to point out these assumptions.

As a tree-hugging college student, then as an environmental lawyer, I heard peers and colleagues decry dismal cattle-catalyzed ecological destruction. I fully accepted these ideas. The accusations have always been especially compelling when accompanied by vivid photos and videos of barren lands out west and Brazilian forests being cut and burned, apparently to make way for cattle. Such photos played a decisive role in my own choice to give up beef—the first meat I swore off—during my freshman year of college.

In this book, I will address many common criticisms of beef, beginning with the burning issue most often cited today: climate change. It is the great environmental challenge of our times. Thankfully, skepticism of human-caused global warming among Americans, including farmers and ranchers, is waning. Yet media coverage continues to sensationalize the matter by brandishing sexy (yet silly) questions like: *Which contributes more to climate change—eating a hamburger or driving a Hummer?* Such articles glibly conclude the hamburger is worse, closing by suggesting that swearing off beef will have more impact than purchasing a Prius.

British journalist, farmer, and former editor at *The Ecologist* Simon Fairlie dates the rhetorical shift to 2007, since which time he notes that climate change "has become the main argument against carnivory." I've made the same observation on this side of the Atlantic.

Like Fairlie, I consider a single document largely responsible. In late 2006, the Food and Agriculture Organization (FAO), a division of the United Nations, issued a report titled *Livestock's Long Shadow* that blamed meat for 18 percent of human-caused greenhouse gases. Cattle took the lion's share of the blame. UN agencies release scores of reports every year, with most scarcely garnering public notice. Yet something about this one proved irresistible to the media—particularly the press release's catchy headline, which suggested livestock cause more global warming than the entire transportation sector.[1] The report's authors later retracted this statement, acknowledging it was a calculation error.[2] But no matter. By the following year, everyone from animal rights and environmental groups to *The New York Times* editorial board was treating the report's 18 percent figure and the press release's spurious claim as gospel truth.

The earth has undoubtedly been subjected to a great deal of improper cattle grazing (far more than most ranchers like to acknowledge). And cattle raising has climate change implications. But not in the ways or to the degree people now tend to believe. Regrettably, the flurry of attention to the cattle–climate connection following the FAO report has shed little light on the issue while generating a lot of heat. In fact, these issues are poorly understood not just by the general public, but also by environmentalists and animal activists, and even within the cattle and beef industries themselves.

FAO's *Livestock's Long Shadow* report only deepened misperceptions. I found myself frustrated with widespread misuse of its figures, prompting me to write an essay titled "The Carnivore's Dilemma," published by *The New York Times* in October 2009.[3] Nothing about livestock is *inherently* damaging to the environment, I argued. The problem lies instead with today's methods of raising them.

In the decade since writing the essay, I've kept my eye on the emerging science of global warming, particularly the role of agriculture and food production. I have amassed a pile of articles, studies, and real-world examples showing well-raised cattle benefiting from soil biology, biodiversity, water, and climate. I've seen ecosystems where animal impact has kick-started soil biology, returning life, revitalizing deadened and desertified lands, even resurrecting long-disappeared watercourses.

My research has led me to conclude that climate change charges against cattle are a red herring. They are not just overblown, they're dangerous. Focusing on cattle and beef is an enormous distraction. It's diverting energy and attention from global warming's main drivers and what needs to be done to address them. Only by cutting fossil fuel emissions and regenerating drying, dying lands can we ever hope to effectively address the looming climate crisis.

To make sense of livestock's true role, we must get beyond sound bites and click-bait. The real story of livestock and climate is complex and nuanced. To untangle this, let's first consider the three main greenhouse gases caused by agriculture: carbon dioxide (CO_2), methane (CH_4), and nitrous oxide (N_2O).

As the biggest driver of climate change, carbon dioxide rightly gets the most attention. CO_2 makes up 82 percent of all US greenhouse gas emissions.[4] Its warming effect lasts hundreds of thousands of years. Essentially, CO_2's warming effect lasts forever. The United States is responsible for a disproportionate share of the globe's CO_2: We have just 4 percent of the world's population but we annually emit 15 percent of the globe's human-caused CO_2. The vast majority of it, 92 percent, comes from fossil fuels. On a per-trip basis, airplanes are pollution-intensive. But 58 percent of US transportation emissions come from personal vehicles—our cars, pickups, minivans, and SUVs. We Americans adore our cars, and we drive a lot.

Agriculture emits far less carbon dioxide than other sectors of the economy. It also has a more complex connection to carbon. Today 14 percent of agriculture-related greenhouse gas emissions are CO_2. Over the long arch of history, farming's carbon emissions came from razing trees, burning vegetation, and plowing ground. Carbon is the stuff of life. Vast amounts of it, held in living tissues of trees and plants, and bound to soils, were released to the air as humans spread across the globe and converted wild areas to cultivated fields. Just how much carbon loss this caused has been difficult to measure. A 2017 study in *Proceedings of National Academy of Sciences* calculated that agriculture has collectively caused the loss of 133 gigatons of organic carbon from the soils, a hefty chunk of total human-caused carbon emissions.[5] By comparison, *all* human activities emitted 737 gigatons of carbon from 1751 to 1987.

This mass migration of carbon from earth to sky has had dire consequences on both ends. The *presence* of CO_2 in the atmosphere has been the main driver of climate change. Meanwhile, the *absence* of carbon in the ground has slowly but steadily drained life from the earth's lands. "We have lost the biological function of soils," explains Barron J. Orr, lead scientist for the UN Convention to Combat Desertification.[6] "If all the land farmed around the world was in no-till, we could probably reverse climate change," Randall Reeder, retired Ohio State University Extension agricultural engineer, has opined.[7]

In modern US farming, most carbon dioxide emissions come from fuels burned to operate tractors, combines, harvesters, automated feeders, and other mechanized equipment. Global agricultural carbon dioxide emissions, on the other hand, are mostly from the developing world, due to land-use changes, especially the razing and burning of forests. During the 1990s, tropical deforestation in Brazil, India, Indonesia, Sudan, and other developing countries caused 15 to 35 percent (depending on the year) of annual global fossil fuel emissions.

Environmental and vegan advocates are rightly alarmed by this clear-cutting, a humanitarian and ecological travesty. But they are wrong to pin the blame on cattle. Cattle are merely instruments under Brazilian law to gain control of land, according to several knowledgeable people I've interviewed. Wealthy landholders seek to push indigenous peoples off wild areas, placing cattle there to claim title. Brazil's controversial president, Jair Bolsonaro, has repeatedly indicated he regards rain forests as lost economic opportunity. He has infamously turned a blind eye as powerful individuals and corporations have ramped up this sort of displacement of indigenous peoples in recent years.

Sue Branford, a seasoned British journalist who's reported on Brazil for decades for the *Financial Times* and the BBC, says the drivers of deforestation are complex. She sees multiple drivers behind it, and agrees that ranching has become something of a "scapecow." "The final objective is soya, often cultivated in conjunction with cotton—both cash crops for export," she explains. "Ranching is generally a step on the way, after logging and before soya."[8]

Brazil has seen huge swaths of tropical forest converted to soybean fields, the most profitable land use. In Brazil's Mato Grosso state, soybeans are grown on nearly 70 percent of newly cleared areas. Over half of the nation's soy harvest is controlled by a handful of international agribusiness companies, which ship it worldwide for animal feed and other food products. Where forests are cut down to make way for soy fields, carbon dioxide is emitted first from burning the trees and, later, from plowing, planting, harvesting, and transportation of the crops grown there.

Of course, no one eating beef need be part of this scenario. Beef cattle throughout the world are mostly kept on grass. The international soy market supplies industrial animal feeding operations, especially for chicken and pork, in Latin America, Europe, China, and, to a small extent, the United States. US cattle feeding get less than 1 percent of its soy from Brazil. In recent years, a growing number of American farmers and ranchers have returned to raising their cattle entirely on grass. They buy no soy, thus having no responsibility for soy-related emissions. It is both inaccurate and unfair to impute deforestation emissions to grass-raised cattle or those fed entirely with locally grown feeds.

This is a fundamental flaw in FAO's *Livestock's Long Shadow* report as well as others that have followed. They simply lump all entities together and pin deforestation emissions on an aggregated meat industry. The result—equally blaming livestock farmers and ranchers the world over for the effects of deforestation occurring in very specific regions with poor land-use policies—is absurd.

Some academics opine that buying beef in the United States contributes to deforestation in Brazil by driving global demand. However, I have never seen any empirical proof for this assertion. It's actually more logical that precisely the opposite is true: By buying beef from US sources, especially grass-based regenerative ranches, we bolster domestic supply channels and effectively reduce pressure on the Amazon. As consumers, each of us can select domestically raised meat. Better still, by carefully choosing our beef from known sources we can directly support farms and ranches using practices that regenerate land and return carbon to soils.

Ironically, people who entirely eschew beef may find it harder to avoid having their purchasing dollars funneled toward deforestation and other destructive practices, like mass monocultures, and toxic pesticides and herbicides. Parts of soybeans go in animal feeds. The beans are also crushed for their oil, over a third of their value, and turned into soy isolate and soy lecithin, all of which are common processed food additives, especially in vegan foods.[9] According to Organic Consumers Association, soybeans from Brazil's deforested areas are also found (unlabeled) in tofu and soy milk sold in American supermarkets.

Mechanized facilities that confine animals around the clock—mostly dairy, pork, and poultry operations—necessarily cause carbon dioxide emissions. These buildings have automated systems for feeding, lighting, sewage flushing, ventilation, heating, and cooling, all of which generate CO_2. These operations, as well as beef cattle feedlots, must also continually provide feed for their animals. The growing, harvesting, drying, and transporting of feeds generate additional CO_2 emissions.

By contrast, pastoral livestock keepers around the world cause negligible carbon dioxide emissions. They have little mechanization; nor do they generally grow or buy feed. For all cattle, but especially those kept on grass, CO_2 emissions are minor. This is obvious when deforestation happening in distant lands is removed from the equation. On top of that, recent research shows that well-managed cattle grazing can actually *add* more carbon to soils than the total greenhouse effect of the beef created, thus having a *net drawdown* effect. (More detail on this later.)

Next, let's consider methane, CH_4, believed to make up 40 percent of US agriculture's current greenhouse gas emissions.[10] As with CO_2, the world situation differs considerably from the domestic. Global methane emissions rose in the final decades of the 20th century, leveled off for the first decade of the 21st, then began rising again after 2008. Climate scholars actively debate the causes of these fluctuations. Notably, though, global methane levels have not run parallel to global ruminant populations.

Journalist Judith Schwartz points out that the lack of correlation between methane levels and the world's ruminant numbers raises doubt about their true contribution to atmospheric methane. She cites a joint 2008 report from the FAO and International Atomic Energy Agency showing that since 1999 atmospheric methane concentrations have been stable, while at the same time the global population of ruminants grew at a rapid rate.[11] Similarly, a 2008 paper by UN agencies, "Belching Ruminants, A Minor Player in Atmospheric Methane," noted "the role of ruminants in greenhouse gases may be less significant than originally thought, with other sources and sinks playing a larger role."[12] The claim that global methane is up because of rising cattle numbers is certainly dubious.

The US methane situation differs considerably from the global one. Environmental Protection Agency figures show US methane emissions steadily and significantly declining in recent decades—down almost 16 percent from 1990 to 2017. Falling US methane levels once again illustrate how methane emissions do not track cattle populations. American cattle numbers had reached a record high in the mid-1980s, then fell precipitously in the latter part of the decade, returning to midcentury levels. Cattle numbers stayed low in the 1990s and 2000s, slowly climbing in the past few years. As with the global situation, there has been no parallel between US cattle numbers and US methane levels. These fluctuations mystify scientists. Knowing what sources of methane are actually causing changes in atmospheric levels is far more complicated than most of us have appreciated.

Another vexing methane question for climate scientists is the impact of shale-oil production, commonly called fracking. The fracking process is known to cause substantial methane emissions. Speaking in response to research suggesting US cattle were causing methane to spike, Dr. Robert Howarth, director of Cornell University's Methane Project, says, "This conclusion is inconsistent with the satellite data." His research reveals that the carbon signature of methane attributed to cattle by some prior research is actually caused by fracking. Yet at the same time total US methane emissions have been falling, there has been dramatically more fracking, according to Dr. Howarth.[13] The fracking riddle again illustrates how difficult it is to accurately assess which sources are generating methane actually contributing to climate issues.

Before diving deeper into cattle-related methane emissions, it's important to first emphasize another serious problem with methane measurements. "The (current) conventional method for accounting for emissions is all wrong for methane," says Dr. Myles Allen. And he would know. Allen is professor of geosystem science in the School of Geography and the Environment, University of Oxford, and head of the Climate Dynamics Group in Oxford's Department of Physics. He has served on the United Nations' Intergovernmental Panel on Climate Change (IPCC) as lead author of "Detection of Climate Change and Attribution of Causes" for the 3rd Assessment in 2001; as review editor on "Global Climate Projections" for the 4th Assessment in 2007; and as lead author on "Detection and Attribution of Climate Change: From Global to Regional" for the 5th Assessment in 2013.

Dr. Allen has also extensively written and spoken about a fundamental flaw in accounting for methane under the current approach, known as Global Warming Potential (GWP). The GWP metric was developed to allow comparisons of the global warming impacts of different gases. It is intended to measure how much energy the emissions of 1 ton of a gas will absorb over a given period of time, relative to the emissions of 1 ton of carbon dioxide (CO_2). Under this method, CO_2 and methane are treated similarly. This is the problem. It ignores the way gases work in the real world. Carbon dioxide lasts "forever, in practical terms," says Allen, so every bit of it builds up in the atmosphere. On the other hand, due to its brief life span and fast natural degeneration, "methane does not accumulate." Thus, while the gases are similar on paper (they are "nominally equivalent," as Allen says) their actual effects on the planet's temperature are very different. And, Allen explains, in contrast with CO_2, "methane emissions don't need to reach zero to stop and reverse methane-induced warming." In other words, the GWP method creates a false equivalence between the two main greenhouse gases. "Methane behaves completely differently than carbon dioxide," says Allen. This failing is crucial to the beef debates—it throws all current estimates of cattle's contributions to global warming into doubt.

Dr. Allen illustrates the failings of the current GWP system with the following specific example. Under GWP, if you started with 1 tCH_4, and it increased at an annual rate of 25 percent for 30 years, this would be converted under GWP to 980 tCO_2 (which has a warming impact); if you started with 1 tCH_4 and there was a gently falling rate, declining 10 percent over 30 years, this would be converted under GWP to 800 tCO_2 (which also has a warming impact); and if you had a steeper methane decline—25 percent over 30 years—that would be converted under GWP to 735 tCO_2 (which also has a warming impact). But the *real-world impact* of each of these situations would be totally different. The first scenario (25 percent methane increase) would contribute to warming. But the second scenario (gentle methane descent) would actually be stable, and the methane *would not actually contribute to warming at all*. Finally, the third scenario (25 percent reduction of methane) would actually be contributing to *cooling*! This shows just how seriously flawed the current accounting system really is.

Fortunately, Dr. Allen has figured out how to correct the accounting system. His corrected accounting system—which he calls GWP*—actually *calculates the warming effect* of the emissions. Under GWP*, the first situation (25 percent methane increase) should be converted to 945 tCO_2; the second situation (10 percent methane decrease) should be converted to 0 tCO_2; and the third scenario (25 percent methane decrease) should be converted to -420 tCO_2. Yes, you read that right—*negative* 420 tons of carbon dioxide. Clearly, the accuracy of these accounting systems really do matter. Dr. Allen's GWP* shows the actual global warming effect of each of these scenarios.[14]

I had the pleasure of listening to a talk by Dr. Allen, then personally meeting him at a Sustainable Food Trust conference in England in July 2019. He told me in conversation that many scientists agree with his criticisms of the current system ("the science here is not controversial at all," he said). But some of those same scientists simply feel changing current methane accounting methods would be too burdensome ("the policy implications are huge"). He also shared with me his own frustration that cattle are the focus of so much climate activism. "It takes attention away from the real problem," he told me, "which is fossil fuels." I could not agree more.

Now let's get back more specifically to food and farming. Methane emissions from ruminant animals, especially beef and dairy cattle, have been repeatedly singled out for criticism. But other parts of the food system also generate significant methane. Wetland rice fields, for example, caused as much as 29 percent of the world's total human-generated methane in the late 20th century. Although some experts dispute it, the rice industry is now generally credited with having significantly reduced its emissions. Either way, rice production today continues to generate a lot of CH_4. The general consensus is that rice farming currently creates about 10 percent of the world's human-caused methane.[15]

In animal farming, liquefied manure storage lagoons are another source of methane. Industrial pork, egg, and dairy facilities often add water to manure (for ease of transport) and hold millions of gallons in storage beneath confinement buildings and in waste storage ponds. This problem is not inherent to animal farming; it is uniquely and specifically associated with large-scale, industrialized agriculture. Neither smaller-scale nor grass-based operations use manure lagoons. "Before the 1970s, methane emissions from manure were minimal because the majority of livestock farms in the U.S. were small operations where animals deposited manure in pastures and corrals," US Environmental Protection Agency documents state.[16] EPA research shows methane emissions from livestock farming jumped with the rapid rise of industrial animal confinement operations, as liquefied manure systems became the norm. Cattle ranches and beef feedlots, however, are not responsible for these emissions; they do not use liquid manure lagoons.

Beef's critics often focus on cattle as prime culprits in methane emissions due to their enteric processes. This seems logical since cattle digestion releases methane. But for several reasons, it's more accurate to consider digestive fermentation from farm animals as something entirely different from CO_2 or fugitive methane emissions from industrial sources like fracking.

First, as mentioned previously, methane is very short-lived in the atmosphere. Carbon dioxide and other warming gases persist for hundreds of thousands of years. For those gases, "The more we emit, the more accumulates in the air," explains UC Davis animal emissions expert Frank Mitloehner. "These gases are called stock gases because they always add up; they don't go down."[17] Methane, in contrast, has a brief atmospheric life span of 10 to 12 years. Its presence in the atmosphere is only fleeting due to hydroxyl oxidation, the natural process that breaks down methane within a few years of being emitted.[18]

Second, CH_4 from cattle is part of earth's *biogenic carbon cycle*. In contrast with industrial sources, this type of methane is not actually *adding* carbon to the earth's ecosystem. This carbon is moving through plants, animals, soils, and air. "As part of the biogenic carbon cycle, plants absorb carbon dioxide, and through the process of photosynthesis, they harness the energy of the sun to produce carbohydrates such as cellulose," a UC Davis publication explains. Indigestible by humans, cellulose is broken down in cattle rumens. A portion of the carbon in cellulose is emitted as methane (CH_4), then converted into carbon dioxide through hydroxyl oxidation. "That carbon is the same carbon that was in the air prior to being consumed by an animal. It is recycled carbon."[19] Said another way, the carbon of ruminant emissions is *old carbon*—carbon already naturally cycling through ecosystems.[20]

Ruminants have been grazing, chewing, digesting, burping, and farting for tens of millions of years. As with millions of wild ruminants before them, carbon released by cattle as methane comes from the air, is taken up by plants and soils, feeds animals, and returns to the atmosphere, where it is available again to power plant growth. This is not pollution; it's the earth's age-old natural carbon cycle.

In North America, for millennia prior to European colonization, grazing herds blanketed the continent, including millions of caribou and deer, an estimated 10 million elk, and somewhere between 30 and 75 million American bison. "The moving multitude . . . darkened the whole plains," Lewis and Clark wrote of bison in 1806. Pennsylvania State researchers concluded that the total number of large ruminants in prehistoric times was larger than the cattle herd in the United States today.[21]

Fossil fuel emissions, in sharp contrast, are not part of this life-supporting cycle. For energy generation, carbon compounds safely tucked away deep in the ground are mechanically extracted then burned. The process causes CO_2 and CH_4 releases all along the way. This carbon has not been powering natural systems by cycling through plants, soils, animals, atmosphere, and plants again. This is *new carbon*. This migration of carbon from earth to sky is what's warming the planet.

Viewed in the larger context, enteric emissions from cattle, from their unique, ruminant digestive system, should be regarded as a single side of a double-edged sword. Allow me to elaborate.

If a human ate a handful of grass, it might make her feel mildly nauseated but would simply pass through her single-chambered stomach, providing little, if any, nourishment. The human digestive system is such that we need to be directly nourished by our food. A cow, however, is nourished by a multi-step process involving an array of tiny living things. A complex of microorganisms—bacteria, protozoans, and fungi—live symbiotically with the cow in her four-chambered stomach. The digestive process involves fermentation in a chamber called a *rumen*, which is why cattle are called *ruminants*. Bovine gut flora are nothing short of miraculous: Billions of tiny organisms foster digestion in such a way that cattle can survive entirely on low-nourishment cellulosic plants like grass. This capacity transforms the globe's vast grasslands into an invaluable component of the human food system.

As a by-product of this unique digestive process, methane is burped (mostly), breathed out (somewhat), and farted (a little) by cattle. If you've ever stood near a cow when she masticates, you cannot help but observe her cud chewing. Methane results from the biological decomposition of the vegetation. This would occur eventually, it should be noted, whether or not cattle were consuming it. Cattle chewing and digesting grass merely hastens the decay. Whenever methane from cattle is discussed, this duality should be kept in mind.

Moreover, although the digestive processes of cattle do generate methane, the causes of these emissions are now fairly well understood, and there are reasonable and increasingly available ways to curtail them. Methane production is aggravated when livestock eat poor-quality forages, throwing their digestive systems out of balance. (For this very reason, FAO climate reports show that cattle in the United States, which tend to have access to better forages, generate notably lower emissions than those in most other parts of the globe.[22]) Livestock nutrition experts have demonstrated that by making minor improvements in animal diets (such as by providing nutrient-laden salt licks), they can cut enteric methane by half.[23] Other practices, like adding certain proteins to ruminant diets, can reduce methane production per unit of milk or meat by a factor of six, according to research at Australia's University of New England.[24]

One particularly promising line of research is seaweed as a feed supplement. Studies conducted in Ireland, Australia, and the United States have shown large reductions in cattle methane emissions when small amounts of seaweed are added to their feed. In 2016, Australian researchers identified a particular seaweed strand called *Asparagopsis taxiformis*. In lab models, the team found grass feed with 2 percent seaweed could cut methane emissions by nearly 99 percent. In 2019, another group of researchers reduced methane releases 95 percent by supplementing a typical US dairy cow diet with just 5 percent *A. taxiformis*.[25]

Myriad other practices have also been found to reduce methane emissions from cattle. Research at University of Louisiana has demonstrated enteric methane emissions can be notably cut when cattle are regularly rotated onto fresh pastures. Other university research has shown that breeding and feeding can lower enteric methane emissions, and that novel techniques like adding small weights to the rumen can reduce methane emissions by nearly a third. A Texas-based biotech company believes it can cut ruminant emissions by 50 percent by giving cattle certain probiotics; others are exploring vaccines that alter the animals' microflora to reduce methane emissions.[26] Research has even shown the presence of dung beetles in meadows can significantly reduce the methane emitted from cattle on grass.[27]

Another interesting line of research relates to methane-consuming microbes. Termites are like tiny cattle. The textbook *Biology of Termites: A Modern Synthesis* describes how the insects live symbiotically with a complex of diverse microbes in their guts that allow them to get nutrition from cellulose.[28] As with ruminant digestion, this process emits methane. So termite mounds were once considered noteworthy sources of global methane. But as the termite textbook notes, research discovered "methane oxidation in mound material and surrounding soil." Field trials have revealed that methane from termite mounds fluctuates significantly based on various conditions, such as temperature. But overall, the mounds produced far less methane than once believed.[29] This is because bacteria living in mounds alongside termites and nearby soils actually consume much of the methane released from termite digestion. The net effect is to cancel out much of the methane emissions.

These methane-oxidizing bacteria (called MOBs or methanotrophs) are a unique group of aerobic, gram-negative bacteria that use methane as their sole source of energy. "They are ubiquitous in nature and, as the major biological sink for the greenhouse gas methane, they are involved in the mitigation of global warming," according to an article in the journal *Applied Microbiology and Ecology*.[30] This has led to studies of grazing systems, including by Professor Mark Adams at Sydney University, in Australia, and Professor Ed Bork of the University of Alberta, in Canada, both of which found portions of enteric methane being consumed by soil-dwelling methanotrophs. A multi-year study by researchers from the School of Ecology and Environment at Inner Mongolia University, published in 2019, studied CH_4 flux and the soil methane-oxidizing bacteria abundance in a typical steppe under grazing, mowing and fencing management in central Inner Mongolia. Researchers measured, over five years, CH_4 flux, plant biomass, and soil physicochemical properties. They concluded: "The results showed that moderate grazing significantly enhanced CH_4 uptake rate and the methane-oxidizing bacteria abundance."[31] Similar results were published in the journal *Biology and Fertility of Soils* in 2020. That study focused on the metabolically active, aerobic methanotrophs in grasslands with three different levels of grazing (light, medium, and heavy). They found that "light and intermediate grazing stimulated the growth and activity of methanotrophs." Notably, the research did not find this effect from "heavily grazed" areas. But this may be an example of poor management.[32]

Much remains to be learned. But to me, these diverse and promising lines of research clearly demonstrate bovine methane emissions are not an intractable problem. The third major greenhouse gas generated by agriculture is nitrous oxide. As with carbon dioxide, the connection to cattle is weak. Nitrous oxide makes up around 5 percent of the total US greenhouse gases.[33] Of that, 73 percent is attributed to agriculture.[34] Almost 90 percent of agriculture's nitrous oxide emissions result from use of manmade fertilizers.[35] The compound is produced in the soil by two microbial processes: nitrification (ammonia oxidation) and denitrification (nitrate reduction). Research from the University of California–Berkeley in 2012, using a nitrogen isotope, proved for the first time that manmade fertilizers stimulate microbes in the soil to convert nitrogen to nitrous oxide at a faster-than-normal rate.[36]

Such research is damning to conventional, chemical-based agriculture, but it is not a strike against beef. It is entirely possible to raise cattle (our own ranch is living proof) using few or no agricultural chemicals, and more and more people are doing just that. Cattle that are neither fed crops grown with commercial fertilizer nor grazed on pastures where manmade fertilizers were applied have no connection with this chain of events.

In fact, studies like the UC Berkeley research actually provide further support for the importance of cattle and other livestock in the food system. Animals offer the best hope for improving the biological life and fertility of soils without chemicals. These methods include grazing animals on grass planted as part of crop rotations, grazing on crop residues, and applying animal manures to croplands.[37]

Now I want to take a moment to dive a little deeper into *Livestock's Long Shadow*'s 18 percent figure for meat's contribution to global warming. It has some serious credibility problems.

For starters, this figure was always an outlier. As Simon Fairlie points out in his thoughtful and meticulously researched book, *Meat: A Benign Extravagance*, the 18 percent number far exceeded most other estimates from reputable scientific organizations. Among those were the Nobel Prize–winning Intergovernmental Panel on Climate Change (IPCC), which said at the same time that the *whole of agriculture* causes between 10 and 12 percent of global greenhouse gases.[38] Likewise, here in the United States, official Environmental Protection Agency reports show US agriculture causing just 8 percent of US global warming emissions.[39] It bears restating this is for *all* US agriculture, not just meat production. (By comparison, EPA reports that US transportation causes 28 percent of emissions.) Clearly, the *Long Shadow* figure never reflected a scientific consensus, and has limited application to animal farming here in the United States where fossil fuel emissions are a far more urgent problem.

Additionally, the 18 percent figure cannot be regarded as objective, scientific data. It is clear *Livestock's Long Shadow*, like other reports, was crafted to support a particular policy agenda. In this case, the recommendations strongly suggest it was devised to strengthen the following argument: The global demand for meat is rising; confined pig and poultry operations have a lower climate change impact than cattle; thus, the world food supply should move away from grazing animals and toward industrial poultry and pork.[40] (Remember, this is the report's perspective, not my own.) Even before its release, *Long Shadow*'s lead author, Henning Steinfeld, a German agricultural economist, was on record making precisely this argument.[41]

I've personally met Steinfeld twice: once when he visited our California ranch, and once on a panel at a livestock conference in Bonn, Germany. From our direct conversations, as well as from listening to him present his views at the conference, it was clear Steinfeld favors just such an approach. He believes that to fulfill a growing appetite for meat, the world should increase confined pork and poultry operations while shrinking herds of grazing animals.

I'm not suggesting Steinfeld is anything less than competent and sincere; he strikes me as both. But authors do have perspectives. And understanding an author's point of view helps explain why certain things are included in a calculation while others are omitted. Such decisions are hotly debated in emerging fields like climate change. Henning Steinfeld views grazing animals as problematic, poultry and pork confinement operations less so. For reasons that will become clear in this book, I strongly disagree.

More recently, *Livestock's Long Shadow*'s 18 percent figure has fallen out of favor within the United Nations, even at FAO. In September 2013, FAO released a report it characterized as an "update" to *Livestock's Long Shadow*. It revised the 18 percent figure to 14 percent, a 22 percent reduction. Similarly, in November 2013, FAO's sister agency, the United Nations Environment Program, released a report called *The Emissions Gap Report 2013*, which stated that *the whole of agriculture* caused 11 percent of greenhouse gases.[42] These and other more recent analyses should take precedence.

The 18 percent figure should never be cited again.

In addition to FAO, others have attempted to quantify livestock's total global warming impact, generally with less credibility. Among them were two policy analysts (neither a climate scientist) who asserted in a 2009 *World Watch* publication that livestock were responsible for a whopping 51 percent of all global warming gases.[43] Predictably, this bold pronouncement attracted media attention.

However, the wild claim quickly crumpled under scrutiny, as it turned out the number was simply hatched by the authors. In interviews, they acknowledged the 51 percent figure resulted from neither new research nor even new anecdotal information. Rather, they explained, they had read an article by a physicist who suggested including livestock respiration in greenhouse gas calculations. The authors (the senior of whom, now deceased, was an outspoken vegan advocate) then simply took the figures from *Livestock's Long Shadow* and recalculated them, adding in a (very large) number for animal respiration. Mind you, the physicist's article was also not reporting results of any fieldwork or climate change research; it was simply a two-page essay in which he espoused an idea. Hmm.

Notably, FAO's newer study (which concludes livestock's role in greenhouse gases totals 14 percent rather than 18) *explicitly excluded* respiration. Hopefully it puts to rest once and for all this breathy idea. This must have shocked the authors of the *World Watch* piece, who had earlier stated: "We heard recently that FAO have been sparked by our work to do their own recalculation . . . They have a lot of money and a lot of people; a lot of good mathematicians, and they have the world's best database on all these things, and so they're going to recalculate their own work and I'm sure that their 18 percent will move towards our 51 percent, or even exceed it."[44] Um, no.

As one would expect, the figures in *Livestock's Long Shadow* have also been dissected and criticized by those who feel they overstate livestock emissions. The most important of these objections, discussed earlier, is that nearly half of the 18 percent number was due to carbon releases from deforestation in the developing world, particularly Brazil, India, Indonesia, and Sudan. The idea of including such emissions was novel, and it's the main reason *Long Shadow*'s number was so much higher than any other previous credible calculation. Yet including deforestation emissions raises several red flags.

In addition to the problems already highlighted, there's this one: A figure for numbers of acres cleared cannot be the basis for a credible annual figure of greenhouse gas emissions. As trees are felled and burned, carbon that was stored in them is released to the air as CO_2. This is a onetime event. Once an area has been cleared it cannot be deforested again the next year, let alone the next. The CO_2 emitted from deforestation happens once, period. To treat this as an annual, recurring figure is patently misleading. The fallacy in taking this approach is illustrated by FAO's own updated figure (14 percent), which was, its authors said, considerably lower primarily because Brazil was taking serious measures to rein in its deforestation problem (although the 14 percent figure, too, includes the deforestation emissions in its total).

Cutting down forests in the developing world, as troubling as it may be, actually has almost no connection to American beef consumption. In theory, soy from deforested areas could be imported to the United States as livestock feed. In reality, however, the amount is tiny because nearly all soy used in US feeds is domestically grown. Less than one-third of 1 percent of all soybeans used as feed in the United States are imported from *all* other countries. Three-quarters of Brazil's soybeans go to China, with nearly all of the rest going to the EU.[45] If any soy at all goes from deforested Brazilian fields to American beef feedlots, it is statistically zero.

By the same token, less than a fifth of all beef and veal consumed by Americans (16 percent, according to USDA researchers[46]) is imported. And over 80 percent of that 16 percent comes from Canada, Australia, and New Zealand. In other words, the most that could possibly originate from all deforested developing countries combined is just 3 percent. All of which means the environmental advocacy pamphlet I saw as a college student depicting my hamburger destroying the Amazonian rain forests was simply a fiction. Deforested lands in the developing world are the source neither of US cattle feed nor of American beef.

Meanwhile, beef isn't causing deforestation in the United States. Trees are not being cut down here to make way for cattle, and, despite a popular myth to the contrary, they never were. Historically, forests in this part of the world were cleared for crop agriculture, timber, and railroads, not for grazing. Vast stretches of the United States—particularly in the Southeast, Midwest, and Far West—were grasslands when Europeans arrived. There is active discourse over the contribution of Native Americans to the creation and maintenance of these grasslands. Regardless of any human role, land available for grazing was in place when Europeans began populating North America with their cattle in the 16th and 17th centuries. These existing grassy areas served as the primary grazing areas for both wild herbivores and domesticated livestock.[47] When humans cleared land of trees or actively prevented foresting, it was done for reasons other than grazing cattle.

I want to flesh out a bit how Native Americans interacted with the land, because it's important to the broader questions of what North American lands looked like and how they functioned before cattle. A broad consensus now exists that Native Americans actively engaged in land management, including extensive burning, seed sowing, and horticulture. This is known from archaeological evidence and Native American oral traditions.

In his book on managed vegetation burning, University of California–Berkeley forestry professor Harold Biswell notes this is also borne out by firsthand observations of various explorers and naturalists.[48] Reasons for burning included the following: improving forage for wild grazing animals and hunting conditions; enhancing visibility for self-defense; facilitating travel; catching other animals, including grasshoppers, lizards, and snakes; improving seed production; clearing brush and undergrowth; and reducing fuels and fire hazards near settlements.[49]

In California, burning was also used to ensure an ample supply of acorns. Food sources were plentiful, and Indians found little need for agriculture. Acorns were their principal plant food, and oak trees were abundant over nearly all of California. "Oak tree preservation was perhaps one of the reasons for the Indians' use of fire," Biswell argues. "Not only do oaks show high resistance to surface fires, but their reproduction is favored by fire. The Indians probably knew this."[50]

Charles Mann's deeply researched book about the Americas before the arrival of Columbus, *1491*, describes how humans throughout the American continents made regular use of controlled burns. "[F]rom the Atlantic to the Pacific, from Hudson's Bay to the Rio Grande, the Haudenosaunee and almost every other Indian group shaped their environment, at least in part, by fire." In addition to the reasons previously mentioned, Mann argues Native Americans engaged in "constant burning of undergrowth" that resulted in increased numbers of herbivores and the predators that ate them. "Indians retooled ecosystems to encourage elk, deer, and bear," Mann writes. He also points to historical evidence Native Americans deliberately vastly expanded the grazing range of the American bison to include areas from New York to Georgia.[51]

In his book *Dirt: The Erosion of Civilizations*, earth and space sciences professor David Montgomery highlights the importance of active land management by early peoples in many parts of the globe. "Long before the last glacial advance, people around the world burned forest patches to maintain forage for game or to favor edible plants. Shaping their world to suit their needs, our hunting and gathering ancestors were not passive inhabitants of the landscape."[52]

When large numbers of Europeans came, they cleared American forests for lumber and railroads, but mostly to grow crops. Livestock's use of the land was always secondary. On some farms, livestock grazed in rotations with crops or on land originally cleared for crop cultivation. Montgomery documents widespread poor land stewardship as Europeans landed in the East and gradually migrated west. Land was cleared and plowed while crops were continuously planted (on southern plantations, often large cotton fields) without properly returning nutrients to the soils through cover crops, rotational grazing, or animal manures.

Such poorly managed fields became bereft of nutrients and severely eroded, eventually rendering them too depleted for crop cultivation.[53] Livestock grazing followed as the default use for such damaged ground. This has given many people the erroneous impression that grazing caused ecological injury that was actually caused by poorly managed crop farming.

Government records show at the start of European settlement, about 46 percent of the land that would become the United States was forested. By around 1900, it had been reduced to 34 percent. However, since that time some areas have reforested, making the overall number quite stable. Today, 33 percent of US land is forested.[54] For decades, urban development—not agriculture, and certainly not livestock farming—has been the primary threat to US forests.

In the past several decades, the United States as a whole has actually seen forest expansion. Total forested acres in the United States *increased* by 26 percent from 1990 to 2005.[55] Forests and cattle are not mutually exclusive.

There is even growing evidence trees help cattle and cattle help trees. Many US cattle spend part of their lives grazing in forests. Livestock keepers around the world are now showing greater interest in creating highly functioning ecosystems involving cattle herds and trees, a practice called *silvopasture*. With proper management, new research suggests, cattle and trees are mutually beneficial. (Cattle's presence on the land also often enables landowners to afford to keep land from being cleared and converted to strip malls.)

Project Drawdown, a collaborative international effort by scientists and policy makers, created a list of the 80 most effective ways to reverse or mitigate climate change.[56] Silvopasture came in at number nine. The group's review of evidence concluded silvopasture "could reduce CO_2 emissions by over 31 gigatons by 2050 if it were ramped up from its current 351 million acres to 554 million acres worldwide."[57]

In the United States, the Department of Agriculture is funding silvopasture pilot projects like the NC Choices initiative at the Center for Environmental Farming Systems in Raleigh, North Carolina. The project is intended to assist farmers in raising beef cattle and other animals outdoor using some of North Carolina's 11 million acres of small, privately owned woodlots. Ecologist Alan Franzluebbers, who is systematically measuring the impacts of the systems, says using silvopasture has potential to create greater soil organic matter than trees grown for lumber without livestock. A farmer in the program says combining trees with livestock gives him "more fertile soils from manure deposits, in turn making the trees taller and the lumber more plentiful." Combining elements is making more complex systems more similar to the complexity of nature. Better for trees, better for livestock, better for farmers.[58]

The wisdom of combining trees and animals is now being demonstrated in numerous parts of the world, including Colombia. Farmers and ranchers are being encouraged to plant trees in pastures used for cattle grazing. "Cattle raised using silvopastoral techniques can digest the forage more easily and reduce their methane emissions by 20 percent," according to researcher Michael Peters of the International Center for Tropical Agriculture in Cali, Colombia.[59]

This is an important reversal of attitudes. Not long ago, agriculture departments at American universities encouraged ranchers and farmers to remove oaks and other trees, under the mistaken belief it would create more productive environments for cattle.[60] Systems combining cattle and trees are now being shown to be more profitable for ranchers and to enhance carbon sequestration in both trees and in the soils.

Colombian silvopasture research serves as yet another reminder of how dangerous it is to generalize about beef's role in global warming.

Everything in agriculture is site-specific and practice-specific. Keep in mind that nearly half of the *Long Shadow* calculation for meat's impact was from deforestation. Yet I've just shown American beef has virtually no connection to deforestation or its emissions. The bottom line is telling an American to quit eating hamburgers because of deforestation is just plain nonsense.

Focusing narrowly on emissions from ruminants while failing to fully consider their larger role in ecosystems and agriculture is myopic. In response to *Livestock's Long Shadow*, the well-respected nonprofit U.K. Soil Association issued a comprehensive report on capturing and keeping carbon in soils. In it, the association cites a European Commission report concluding that European grasslands have the potential to sequester large amounts of carbon on an ongoing basis. The sequestration of carbon in agricultural grasslands of the U.K. was calculated at around 670 kilograms of carbon per hectare per year. That amount, the report notes, "would offset all the methane emissions of [U.K.] beef cattle and about half those of dairy cattle."

Specifically referencing Henning Steinfeld's preferred policy direction, transitioning from grazing animals to intensively confined pork and poultry, the Soil Association points out the folly of such an approach. "Advocates of a shift from red meat to grain-fed white meat to reduce methane emissions could therefore find that this has the perverse effect of exchanging methane emissions for carbon emissions from soils and the destruction of tropical habitats (to produce soya feed), as well as having a far reaching impact on our countryside, wildlife and animal welfare."[61]

This leads to my final critique of *Long Shadow*'s calculations, also relating to the need to look at the question more holistically. This one is fleshed out by University of California–Davis animal sciences professor Frank Mitloehner, whose research is focused on livestock air emissions.[62] He points out that the calculations fail to account for what he terms "default emissions." "[I]f domesticated livestock were reduced or even eliminated regionally, the question of what 'substitute' would be produced in their place has never been estimated." In other words, if livestock were removed, would all the climate change emissions attributed to them (now 14 percent) be eliminated? Clearly not, because the food and other necessities made from those animals would still need to be produced and would still cause emissions. For instance, the earlier discussion of Brazilian soy makes clear that foods eaten in place of meat will have their own (perhaps worse) environmental consequences. There's no sense in talking about mitigating climate change by reducing or getting rid of a particular source of emissions unless it can likewise be demonstrated what replaces the source would have lower emissions.

Cattle provide a great deal more than meat and milk. As living animals, they give invaluable biologically active organic fertilizer in the form of manure. In many parts of the world, cattle provide irreplaceable power for plowing and transport. Post-slaughter, their hides provide leather, while additional fertilizer is made from their blood and bones. And it goes far beyond that. Modern slaughterhouses and processing plants are exceptionally efficient at putting each part of the animal to use—making everything from pharmaceuticals, to blood vessels used in human transplant procedures, to tennis racquet strings. Cow hooves are turned into a wide variety of important products. For example, a protein called keratin is turned into a special fire-extinguishing foam used by airport fire and rescue teams to put out hotter, high-intensity fires.[63]

"Therefore, to estimate accurately the 'footprint' of all livestock, 'default' emissions for non-livestock substitutes need to be estimated and compared to livestock emissions (e.g., manure versus [chemical] fertilizer, leather versus vinyl, wool versus microfiber, etc.)," Mitloehner points out. Only then can "the net [greenhouse gas] differences between livestock and other land-use forms" be estimated and an accurate accounting of livestock's true climate change impact be calculated. *Livestock's Long Shadow* did not take account of default emissions. Nor has any other report I've seen that attempts to calculate livestock greenhouse gas emissions. This failure is a gaping hole in the argument against meat.

I've lingered a bit here on the calculations for the 18 percent and 14 percent numbers because these figures have been so widely cited. For the reasons I've laid out, I consider both of those emissions figures inflated. But my point here is not to argue for any particular figure. Rather, I want to demonstrate that the matter of meat's connection to climate change, especially for cattle, is far from a "case closed." What's much more important than the precise figure is the overall question of whether or not cattle truly aggravate the world's very real and urgent global warming crisis. And there are good reasons to doubt they actually do, or at least that they *unavoidably* do.

This distinction is critical. Right now, the public discussion is simplistic and clear-cut: "Cattle cause climate change; the solution is to stop raising them and quit eating them." Such a dualistic, simplistic way of thinking actually skirts the crux of the issue. The real question is whether cattle *can* be raised in ways that are neutral or even beneficial to our planet's health. The more I've learned about this topic over the past two decades, the more I believe they can. Truly, "It's not the *cow*, it's the *how*."

Cattle as Climate Change Mitigators

Carbon can be taken from the atmosphere and tucked into the soil. The process of carbon sequestration simultaneously reduces global warming emissions and improves soil health and functionality. Carbon content of soils can be enriched by "working the land with the goal of building topsoil, encouraging the growth of deep-rooted plants, and increasing biodiversity," explains author Judith Schwartz. "[R]ather than focusing on growing crops, the intention is to grow the soil."[64]

Carbon sequestration was entirely left out of both of the FAO calculations we've just examined for meat's global warming contribution. FAO acknowledged its failure to include carbon sequestration, even admitting the omission could be substantial. But it justified this decision by stating it "could not estimate changes in soil carbon stocks under current land use management practices because of the lack of global data bases and models."[65] In other words, FAO is saying it's difficult to accurately quantify carbon sequestration at this moment. Rather than include an estimate, it simply omitted sequestration entirely from the calculations. This is surprising because, at the same time, the 2013 report acknowledged, "[g]rassland carbon sequestration could *significantly offset* emissions."[66] This suggests even FAO's lower figure overstates cattle's role in global warming, perhaps considerably.

FAO's somewhat dismissive treatment of the issue might give the impression sequestering carbon in soil is just the wishful thinking of creative cattle ranchers. Nothing could be further from the truth. "Soil carbon sequestration at a global scale is considered the mechanism responsible for the greatest mitigation potential within the agricultural sector, with an estimated 90 percent contribution to the potential of what is technically feasible," states a 2012 study by an international team of scientists in the *Proceedings of the National Academy of Sciences*.[67] This number has gained acceptance among many scientists. Virtually the same figure was used by scientific advisers to the IPCC, the international body set up by the United Nations Environment Program, which stated, "89% of agriculture's [greenhouse gas] mitigation potential resides in improving soil carbon levels."[68]

Despite its enormous potential for climate change mitigation, policy makers at every level have long ignored opportunities to capture (*re*-capture, actually) carbon in soils. In 2009, the U.K. Soil Association sought to remedy that neglect by undertaking the largest and most comprehensive scientific review of soil carbon ever. Its scientists reviewed 39 comparative studies of organic farming soil carbon levels, from various countries and climates.

The Soil Association's report concludes soils would achieve "high carbon gains" if chemical-based agriculture converted to organic. Animals play a key role in organic farming. The studies revealed farming enterprises that include animal manures, crop rotations, cover cropping, and composting have higher levels of carbon in their soils. More specifically, the report found that organic farming enhances soil carbon levels by putting more organic matter back in soils, and in forms that efficiently produce soil carbon. This is achieved by integrating cropping and livestock systems, which assists formation of soil carbon, and by increasing the proportion of land covered in vegetation, which promotes microorganisms that stabilize soil carbon once formed.[69]

Farm animals, especially grazing animals, the Soil Association says, are essential to maximizing farming's potential for carbon drawdown. "Grass-fed livestock has a critical role to play in minimising carbon emissions from farming and this must be set against the methane emissions from cattle and sheep." Grasslands for livestock grazing, including permanent pastures and temporary grass leys (grass cover crops on mixed farms), hold "vitally important soil carbon stores," the analysis determined. The report also notes plowing up grazing areas for crop production has resulted in the U.K. losing 1.6 million tons of carbon to the atmosphere every year (amounting to an additional 12 percent of the U.K.'s agricultural greenhouse gas emissions).

Globally, transitioning to organic farming would lead to large increases in carbon sequestration. "Organic best practices," the Soil Association concludes, hold the staggering potential for sequestering 1.5 billion tons of carbon per year, an offset of about 11 percent of *all* anthropogenic global greenhouse gases (not just those related to agriculture), for at least the next 20 years. And the report's calculations were intentionally conservative. Not included, for example, was "the increase in agricultural soil carbon storage that would result from the almost certainly greater percentage of farmland that would be in permanent grass with widespread organic farming."

Given its tremendous promise, you may be wondering why we hear so little about soil carbon in policy discussions and mainstream media stories about climate change. Could it be that soil—often thought of as just dirt—is simply not sexy enough to capture the public's attention? In an era when news coverage is driven by a competition for the most clicks, soil has been sorely neglected. Soil erosion, arguably as threatening to our long-term ability to survive as climate change, has long suffered the same fate.

The subject may not make ideal click-bait, but soil is our planet's workhorse. "Soil truly is the skin of the earth—the frontier between geology and biology,"[70] says earth and space sciences professor David Montgomery. "Soil is our most underappreciated, least valued, and yet essential natural resource."[71]

Soil is not only essential to food production but also contains massive amounts of carbon. Losses of carbon from earth's soils have accounted for a full one-tenth of *all* human-caused carbon emissions since 1850. Carbon that remains in soils today is still about *three times* as much as what's in the atmosphere and *five times* as much as what's held in the world's forests. An estimated 2.5 trillion tons of carbon is contained in soil, compared with 800 billion tons in the atmosphere and 560 billion tons in all plant and animal life. Put another way, more carbon resides in soil than in the atmosphere and all plant life *combined*.[72] Even in their depleted state, soils are such gigantic carbon banks just a small percentage increase would have dramatic effects on the climate. It is estimated increasing average global soil carbon levels by just 1 percent would reduce atmospheric CO_2 by 2 percent.[73]

In France, where public support for farming is strong, the government has energetically pursued the opportunity to mitigate climate change and benefit agriculture by focusing on soil carbon. At the COP21 Paris climate summit in 2015, France launched the "4 per 1,000" initiative. "It aims to boost carbon storage in agricultural soils by 0.4% each year to help mitigate climate change and increase food security." According to a 2018 article in the journal *Nature*, the French initiative requires better communication and coordination among scientists, businesses, public and private enterprises, policy makers, and the public. "Soils must be recognized as natural capital that can contribute significantly to national economies and human welfare."[74]

Research completed in June 2020 by collaborating French scientists from various institutions highlighted the value of animal impact in capturing carbon in soils. The scientists measured microbial activity, fertility, and organic matter in land that was mechanically mowed compared with land that was grazed. "Our results," they conclude, "showed higher soil organic carbon (SOC) and nitrogen contents in the surface soil under grazing as compared to mowing." They also found "higher microbial biomass." In addition, the team found more enzyme production per microbial biomass with mowing, which, they explained, leads "to more degraded soil organic matter in the mowing system." This research underscores the importance of grazing animals in France, and elsewhere, as concerted efforts to capture carbon in the soils ramp up.[75]

California is taking similar measures to encourage soil carbon capture. Former governor Jerry Brown set a target for California to be carbon neutral by 2045. As part of the effort to meet the goal, the state adopted the Healthy Soils Initiative. A state report found "farms and forests could absorb as much as 20 percent of California's current level of emissions." Kate Scow, professor of soil microbial ecology at the University of California–Davis, told National Public Radio, "There's great potential for agriculture to play a really important role" in reaching the state's climate goals. Hundreds of farmers have signed up for the program. Many are hoping it will greatly expand it in the future.[76]

The idea of soil as a key part of a global solution to climate change is finally beginning to get its due. Using natural solutions have multiple benefits, such as improving soil productivity and food quality. They are also far more cost-effective than many high-tech approaches. A 2017 report published in the *Proceedings of the National Academy of Sciences* about natural climate solutions concluded, "ecosystems have the potential for large additional climate mitigation by combining enhanced land sinks with reduced emissions." More specifically, the report found, "Natural climate solutions (NCS) can provide 37% of cost-effective CO_2 mitigation needed through 2030 for a >66% chance of holding warming to below 2°C . . . Most NCS actions—if effectively implemented—also offer water filtration, flood buffering, soil health, biodiversity habitat, and enhanced climate resilience."[77]

The most tantalizing part of this story is, unlike emissions from burning fossil fuels, the loss of carbon from soils can be reversed: The soil carbon bank can be restocked and restored. "The principal component of the soil carbon store is humus," the Soil Association explains, "a stable form of organic carbon with an average life-time of hundreds to thousands of years." Soil humus levels also determine how much water is held by soils and how well they drain excess water. "Low soil carbon levels are therefore likely to exacerbate the impacts of climate change, by increasing the risk and severity of droughts, water shortages and surface-water flooding."[78]

The science of soil carbon sequestration is young. Knowledge about soil biology is far more sophisticated today than just a decade ago and is rapidly expanding. The amount of carbon in soils was previously thought to be determined simply by how much organic matter was added. Now it is understood that complex questions of microbiology affect whether organic matter is converted to stable soil carbon. The amount can vary from just a few percent to up to 60 percent.[79]

At the risk of glazing over the reader's eyes, because this is such a crucial point, I'm going to get granular here for a moment about how soils store carbon. Trust me: This is more scintillating than it sounds.

Here's an intriguing place to start: Soil sequestration takes glue, a material that literally tacks carbon to the earth, keeping it safe from washing or floating away. The sticky substance involved is *glomalin*—a protein produced by underground fungi. Under a microscope, glomalin looks like strands of honey wrapped around the roots of plants. In 1996, glomalin was discovered and named by USDA soil scientist Dr. Sara

Wright, assisted by then-grad-student Kristine Nichols, who later carried forward the work. Years of lab experimentation and field research led to Drs. Wright and Nichols documenting that glomalin makes up 15 to 20 percent of soil's organic matter. They also began unraveling glomalin's complex, mysterious functions.

Arbuscular mycorrhizal fungi, which synthesize glomalin, function symbiotically with plants. Formed as networks of fine threads (called *hyphae*), they envelop plant roots and extend their thin fingers out into soils. These fungal filaments are highways for two-way exchanges: They transport nutrients from the soil to the plant as they carry carbon from the plant to the soil. Glomalin is a carbon-based molecule, while mycorrhizal fungi are carbon-based life-forms. Both utterly depend on carbon, which they obtain only from the roots of a living plant.

Nichols refers to carbon as the *currency* in these exchanges. The plant uses carbon to "buy" nutrients from the mycorrhizal fungus. The fungus may then use that carbon to "buy" nutrients from other microorganisms in the soil.

Glomalin "glue" plays several roles in these processes. It coats the fungus's hyphal threads, assisting in nutrient cycling while also protecting them. It also aids in forming and stabilizing clumps, known as soil aggregates, helping plant health and soil structure.[80] (Proteins left behind by earthworm locomotion may also play a role in soil aggregation, although studies on that question are somewhat conflicting.[81])

Soil clump formation may sound terribly banal. But it's a cornerstone of carbon sequestration. In aggregation, soil's mineral particles are clustered together and encapsulated. This stabilizes the carbon-containing decomposed plant and animal matter (humus), and protects it from degradation. Soil aggregation gets carbon into a stable situation and keeps it there. Interestingly, carbon from plant roots has been shown to be especially stable, lasting over twice as long in the soil as carbon from plant stems and leaves.[82]

Healthy soils are full of these tiny clumps. Soil aggregates "provide structure to soil for better water infiltration, water holding capacity, and gas exchange, and increase soil fertility by providing organic carbon (that is, food) to soil organisms, which use this food as energy to release plant nutrients from the soil," Dr. Nichols explains.[83]

Nichols's field research has shown soils under native grasses in North Dakota—for example, switchgrass, blue grama, big bluestem, and Indian grass—have higher levels of glomalin than those planted with non-native grasses. This raises the intriguing possibility restoring native grasslands could be a key component of a global strategy to fight climate change. "The more glomalin in a particular soil, the better that soil probably is," says Nichols. And up to a certain saturation point, the more glomalin, the more potential for carbon sequestration.

Correspondingly, Dr. Nichols's research has also revealed that plowing the ground injures the root-loving fungi and reduces glomalin. The living hyphal networks "are physically torn by tillage," Nichols explains. When tillage disturbs soils, fungi must allocate carbon to rebuilding their filamentous networks rather than either extending the networks or producing glomalin. In contrast, farming with low or no tillage increases mycorrhizal fungi and glomalin. Levels of both are particularly robust in rangelands and in cropping systems with diverse rotations.[84]

Because glomalin and hyphal networks are so important to getting carbon into soils, the greatest opportunities for carbon sequestration lie in grazing areas. This means rangelands and pasturelands, especially those with native grasses, as well as in farming operations where grazing is part of a diverse crop rotation.

To effectively mediate the underground exchanges, the fungus must maintain an extensive hyphal network, known as a mycelium. The more expansive and developed the mycelium, the more interactions it can have with bacteria and other microbes inhabiting the soil. Those microorganisms break down organic matter and make the nutrients available for the kind of carbon-based bartering I've just described.

The transactions have several steps. For example, the fungus feeds bacteria growing on its hyphae some of the carbon it gets from the plant. The bacteria then use this carbon to produce enzymes, proteins that act as catalysts in living organisms. The enzymes then make nutrients available for the plant.

Having a variety of crop rotations rather than simple monocropping aids the functionality of the complex system. It expands the types of organisms living in the soil, which improves nutrient cycling as decomposition and enzymatic reactions release bound-up minerals. The fungi carry these minerals to the plant and exchange them for carbon.

Nichols's research has also determined that synthetic fertilizers, particularly phosphorus, create unfavorable conditions for such transactions. Phosphorus is a key nutrient that mycorrhizal fungi trade for carbon from the plant. When commercial fertilizer is added, the plant gets lazy. Instead of doing the work of bartering, it simply takes nutrients from the fertilizer rather than going through the trouble of exchanging nutrients for carbon. Falling numbers of exchanges takes a toll on soil health. Synthetic nutrients ultimately lead to debilitated mycorrhizal networks and lower glomalin levels.

The discovery of glomalin and its functions are some of many recent revelations about complex, abundant life in soils. There is infinitely more going on beneath our feet than most of us have ever imagined.

Keeping living plants in the ground for as much of the year as possible is critical to powering this bustling underground economy, Nichols told me by email. The Soil Association report likewise emphasizes that the more time vegetation covers the soils, the better.[85] Growing plants conduct photosynthesis. Photosynthesis provides the carbon needed for mycorrhizal fungal growth and glomalin production. Without plants in the ground, the cascade of intricate processes grinds to a halt.

The newly emerging science of glomalin solidifies the importance of grazing animals. It also reinforces the pivotal role they play in well-designed food systems that have environmental sustainability as a primary objective. To maximize glomalin in soils, we need grazing lands. No agricultural lands are more continuously covered with vegetation than well-managed permanent pasture and rangelands. Properly stewarding land to raise healthy cattle herds also means managing it for vibrant underground communities, which in turn fosters the capture and storage of atmospheric carbon.

Grass is uniquely capable of forming soil carbon. The U.K. Soil Association's carbon sequestration report explains that grass's special power for increasing soil aggregation comes from its high root densities, fine root hairs, and high mycorrhizal fungal levels. Animal manures and composts are also particularly valuable in creating soil carbon. These facts make animals a key component of strategies to increase global carbon sequestration.[86]

Cattle grazing ensures the continued existence of grasslands in the United States and other parts of the world. It makes protecting open space from development economically viable for landowners. And because grazing benefits the complex biology of soils, it aids in keeping those lands blanketed in growing vegetation. In fact, properly timed cattle grazing can increase plant cover by as much as 45 percent, North Dakota State University researchers have found.[87] Grazing by large herbivores (cattle included) is actually essential for well-functioning prairie ecosystems, according to research from Kansas State University.[88]

The Kansas-based nonprofit Land Institute concurs. The institute authored a 50-Year Farm Bill that proposed increases in perennial crops and permanent pasture. "We see future herbaceous perennial grain-producing polycultures being managed through fire and grazing, just as the native prairie was 'managed,'" institute founder and president Wes Jackson told me. "The large grazer on grassland has always been an integral part of the system here in North America."[89]

Converting grassland to cropland causes carbon losses, while converting cropland to grass results in soil carbon gains. Soil experts Rattan Lal and B. A. Stewart collaborated in a landmark survey and report, with recommendations for improving the world's soils. Lal is a professor of soil physics in the School of Natural Resources and director of the Carbon Management and Sequestration Center, College of Food, Agricultural, and Environmental Services, at Ohio State University who won the prestigious 2020 World Food Prize. Stewart is a distinguished professor of soil science at the West Texas A&M University, and is a past president of the Soil Science Society of America.

In "Food Security and Soil Quality," Lal and Stewart report that "the potential of fallows [grass rotations to crop land] to increase carbon contents has also been widely documented."[90] They note that due to the destructive effects of tillage on populations of "soil macrofauna" (like insects), soil cultivation "can strongly modify the soil functioning and especially soil carbon storage."[91] They report an especially notable decline in soil organic carbon when natural vegetation is replaced with crops.[92]

Fortunately, some of this lost carbon can be reclaimed. A study published in the journal *Global Change Biology*, which analyzed 74 soil carbon research studies, concluded that merely converting cropland to pasture resulted in a 19 percent increase in soil carbon levels.[93]

Animal grazing maintains grassland ecosystems both above- and belowground.

Skeptics of soil's potential to soak up carbon argue the rates of sequestration tend to diminish within a couple of decades after a switch from cropland to pasture or from improved farming practices. This point is disputed. But regardless, as the Soil Association points out, "it is the next 20 years that will be critical in policy terms for delivering major greenhouse gas reductions." Moreover, the association notes, carbon sequestration continues for as much as 100 years, just at lower rates.[94]

Decades of work in this field have taught Dr. Lal that enormous amounts of carbon can be captured in soils. Carbon in the world's agricultural soils has been depleted between 50 to 70 percent, Lal says. Yet he sees potential for restoration efforts to increase the carbon content of the globe's soils by upward of 1 to 3 billion tons a year, equivalent to approximately 3.5 to 11 billion tons of CO_2 emissions.[95] In other words, capturing carbon in soils could solve as much as one-third of all annual human-generated carbon emissions.[96]

In her book *Cows Save the Planet*, journalist Judith Schwartz explores the work of soil scientists like Dr. Lal as well as ranching practitioners. She builds a credible and compelling case that carbon sequestration triggered by well-managed grazing can be part of an effective strategy to reverse climate change. Among scientists advocating for the important role of grazing animals is Australian soil ecologist Dr. Christine Jones. Her blueprint for building carbon-rich soils includes short pulses of grazing with high densities of animals. With such land management, Jones says, "evidence of new topsoil formation can be seen within 12 months, with quite dramatic effects often observed within three years."[97]

This still leaves the question of soil carbon sequestration's potential magnitude. Is it limited to mitigating some of the greenhouse gases from cattle or does it raise the tantalizing prospect that cattle could actually have a *net positive* impact on climate change? This is where things really get interesting.

To this point, I've been arguing merely cattle are blamed for a disproportionate share of greenhouse gases, and that those gases can be largely offset by multiple beneficial effects of grazing. Now I want to propose a far more radical idea: Cattle are not, in fact, a climate change *problem* at all; instead, cattle are actually among the most practical, cost-effective *solutions* to the warming of the planet. Unthinkable? Perhaps. But this is the very case currently being made on the world stage by wildlife ecologist Allan Savory.

Although not himself a rancher, Savory has been well known for decades in farming and ranching circles for his innovative teachings on planning and grazing, a system he calls Holistic Management Grazing or Holistic Planned Grazing. In recent years, Allan Savory's fame has gone mainstream. This is largely thanks to a March 2013 TED talk (viewed 7.1 million times, at last check) in which he takes the highly controversial position that cattle are the world's single best hope for reversing climate change.

Savory is a true iconoclast, equally offending nearly anyone imaginable: academics, ranchers, conservationists, and policy makers, among others. His advocacy for using the bovine, as he puts it, "as a bulldozer that doesn't use diesel, to improve the land" particularly outrages the legions of environmental and vegan advocates who desperately want to see cattle raising go the way of the butter churn. Meanwhile, his ideas that herds should be dense, not sparse, that compaction is beneficial to the land, and that "native" and "non-native" plant designations are practically meaningless, run directly counter to mainstream rangeland and ecology teachings. He hates feedlots and disagrees vehemently with the way most cattle are being raised in the industrialized world today, so neither the cattle nor the beef sector has much use for him. He has, in other words, no built-in constituency.

Yet Savory and his organizations are simply too compelling and credible to be ignored. For decades, he has tirelessly labored to restore grasslands on five continents. In the United States, he founded and guides the Savory Institute, which works both through seminars and publications and out on the land, teaching and demonstrating land improvement through well-planned grazing. In Africa, he founded the Africa Centre for Holistic Management. He's given major lectures at leading academic institutions including Harvard, UC Berkeley, and Tufts universities. The Savory Institute and he himself have received highly prestigious awards, including the 2010 Buckminster Fuller Challenge award for the organization's work "to solve the world's most pressing problems," and the 2003 Banksia International Award, given to the person doing the most for the environment on a global scale.

Today Savory's methods are being followed in the management of some 40 million acres the world over.[98] Some of the results are breathtaking. I have been on numerous ranches (including our own) that loosely follow Savory's ideas, and have seen photographs and videos from locations in several parts of the world where his ideas are being put more strictly into practice. Arid, denuded areas once classified by ecologists as "beyond restoration" have been transformed into lands rich in water and teeming with plants and animals. Biodiversity has soared.

And astonishingly, cattle are the linchpin.

How is this possible? We all "know" that big, heavy animals—cattle above all—do not improve land, they damage it. They do not add water or vegetation to the land, they diminish it. And surely they don't benefit wild animal populations, they decimate them. Right?

For the first part of his life, Allan Savory says he "knew" all these things, too, and he knew them well. As a youngster growing up in southern Africa, he often visited ecologically damaged regions said to be stressed by too many animals, wild or domesticated. As a university student of wildlife ecology, Savory was taught again and again the incontrovertible truth that soil erosion and desertification around the globe resulted from animals trampling and chopping up land, and, especially, from them grazing too heavily on the vegetation. "I was taught that overgrazing was causing the land to turn to deserts," Savory recounts. But, he continues, "We were once equally certain the world was flat."[99]

"My education began *after* I left university," he says.[100] But when he started his career, Savory accepted what his professors and textbooks had taught him. So certain was he of the doctrines he'd been instructed that when he became a government game officer in Northern Rhodesia (now Zimbabwe), most of his work to restore and protect wild areas involved getting animals off the land. "Resting" land by removing large animals was considered the surest and best land salvage strategy.

As a young wildlife ecologist employed to protect an important game preserve, Savory wrote a report that recommended killing 30,000 elephants. "It was," he recalls, "the hardest thing I ever did. Because I love elephants." The idea of shooting tens of thousands of elephants was not only repugnant to Savory, but also horribly unpopular with the public. For this reason, a peer review of the report by other ecologists was ordered. The reviewing scientists agreed: To restore the land, reducing elephant numbers was essential.

Because of Savory's report, some 20,000 elephants were killed. As horrific as that was, the worst was yet to come. That was when it became apparent that despite the slaughter, the land's condition continued to decline. The gruesome horror of killing tens of thousands of elephants had been for nothing. Savory now refers to his recommendation to cull all those elephants as the biggest mistake of his life.

"We all believed," Savory wrote in 1988 about his earlier work in African game reserves, "that with dramatically fewer animals living there the area would naturally recover, but it continued to deteriorate." Like his colleagues, where cattle were present, he blamed them. "In my days as a young scientist, I hated livestock and the people running them,"[101] he recalls.

Actually, Savory says, tongue in cheek, he later realized "[w]e scientists had the bull by the udder."[102] Again and again, Savory was witnessing reductions in animal numbers resulting in greater environmental collapse.

One good thing did come of these tragic experiences. They set Savory on the path of dedicating his life to reversing desertification through practices that were actually working in the field. Through study and observation, he eventually came to believe that animal impacts are key to properly functioning ecosystems. Winnowing down animal numbers is not the answer, not in the least. Instead, the solution lies in making herds of domesticated animals function more like the wild herds of herbivores with which the ecosystems evolved.

"We can now see that it was not the livestock per se that caused significant damage in the past, but the way we managed them,"[103] Savory says. He decries widespread damage in many parts of the globe from improper livestock management. Yet he now believes that more often than not, land in the American West and elsewhere is not being overgrazed. Rather, it is, in Savory's words, being "over-rested." Although that statement clashes with today's conventional wisdom, the critical importance that *disturbance* on the landscape plays in the diversity, structure, and function of ecological systems has been borne out by numerous scientific studies.[104]

Savory believes desertification results primarily from reduced biodiversity—the loss of the mass and diversity of plant and animal life. He notes that sustaining life in any environment requires maintenance of a basic life cycle of birth, growth, death, and decay, in which nutrients are cycled continuously. In perennially humid environments, decay is fostered by a high population of microorganisms. "As vegetation dies throughout the year, organisms continually break it down."[105]

For areas with less moisture, the cycling of life is more problematic. Savory believes that for those parts of the world, the key question is how "brittle" they are. The concept is distinct from fragility. "[D]etermining the degree of brittleness becomes a prime factor in the management of any environment," Savory wrote in his 1988 book *Holistic Resource Management*.[106] The brittleness of an environment is based on how organic matter decays, and on the order of succession of organisms in the environment.[107] The least brittle environments have 100 percent biological decay processes, whereas chemical (oxidative) and physical (weathering) decay processes become dominant as one moves toward brittleness.[108]

Moisture in the environment is a critical factor. But much more important than total rainfall is "the distribution of precipitation and atmospheric humidity throughout the year," Savory writes.[109] Other factors affecting how brittle an environment is include elevation, temperature, and prevailing winds. "Brittle environments commonly have a long period of non-growth, which can be very arid."[110]

Understanding an area's brittleness is important, according to Savory, because it determines the way land should be managed, which is highly site-specific. "[T]he old belief that all land should be left undisturbed in order to reverse its deterioration has proven wrong," he wrote. "Only non-brittle environments respond this way. In brittle environments, prolonged non-disturbance will lead to further deterioration and instability."[111] Grasslands, he found, could lie anywhere from 1 to 10 on the scale of brittleness.[112]

The most important indicator of grassland health is the amount of bare ground in between plants, according to Savory. The more bare ground, the worse the land's health. Ground without vegetation is exposed to the desiccating sun, the force of wind, and the power of rainfall, which dry, blow, and wash soils away.

Savory emphasizes what he terms *effective rain*. Total rainfall is irrelevant, he says; what matters is rainwater taken up by plants, held in soils, or going into groundwater. *Ineffective rain*, in contrast, is that rainwater that runs off or evaporates.[113] Soils, he notes, are by far the world's greatest reservoir of fresh water.[114]

I first heard about Allan Savory and his unique methods for land restoration more than a decade ago, from people who managed our own ranch and taken courses on holistic land management. Savory's books have always been on our own bookshelves, and his teachings have long served as part of the basis for how our own ranch is run. I've had the pleasure of meeting him and hearing him speak on several occasions. Allan Savory is both credible and inspiring. Keeping cattle fairly densely congregated, in prescribed tracts of land, moving them frequently, and allowing the land to rest for extended periods between grazings are all ideas we've adapted from Savory's teachings.

Yet Savory advocates a much more particular and comprehensive management approach than the one we currently follow. And it's an aspiration of mine that our ranch will continue, over time, to move still closer toward his methods.

Stated succinctly, Savory advocates animals be kept in dense herds and moved often; that grazing stimulates biological activity in the soil; that animal waste adds fertility; that hooves break the soil surface, press in seeds, and push down dead plant matter so it can be acted upon by soil microorganisms; that all of this generates soil carbon, plant carbon, and water retention; and that this is the only way to stop and to reverse desertification the world over.[115] "The actual grazing plan will thus be different on each ranch (in each season) and will be continuously changing, according to conditions on the ground."[116]

The philosophy for Savory's approach is to re-create, to the extent possible, the conditions under which grasslands evolved. He never argues cattle don't alter the land. Quite the reverse. After a long career working as a wildlife ecologist and land restoration adviser, Savory is the first to acknowledge that the presence of cattle changes a landscape's ecology. "Where a cow places her hoof today begins a chain of reactions that ensures that that spot will never be exactly the same again," he states.[117] He classifies the bovine's impacts on the land into three types: compaction ("heavy animals on small hooves"); breaking the land surface; and cycling vegetation faster than if they weren't there.[118]

When cattle are not well managed, these impacts can be extremely damaging. He notes that in most arid and semi-arid regions, humans have long grazed livestock, and usually in ways that harm the land. "When livestock management practices produce bare ground, a critical share of available moisture either evaporates or runs off. Springs dry up; silt chokes dams, rivers, and irrigation ditches; and less water remains for agriculture, industry, and people in nearby cities."[119]

Many ecologists would agree with that statement. But they would attribute the damage to something else: overgrazing and the presence of too many domesticated grazing animals. Here Savory's views diverge sharply from the mainstream. "Overgrazing has nothing to do with animal numbers and everything to do with timing," he urges. "Think of it like your lawn: it makes no difference to the grass if you use one mower or 50 mowers. All that matters is how frequently you mow it."[120]

In fact, Savory says, significant numbers of animals are absolutely essential, and they must be kept in dense herds. Sparse numbers are insufficient to cause desirable disturbances. Trampling and breaking the ground surface, as hooved animals are particularly prone to doing, Savory argues, are actually essential to grassland ecosystem function. Such was the impact that massive herds of wild herbivores (along with the predators that stalked them, whose presence ensured that herbivores stayed closely congregated) had on the earth for millions of years. Grasslands evolved under such conditions. Stated another way, the plants and microorganisms of grassland vegetation co-evolved with animals not just to *tolerate*, but to actually *need* grazing.

Here's how Savory himself explains it:

[M]ost of these perennial grass plants developed alongside vast herding herbivores. The animals were dependent on the grass for food, and the grass was dependent on the animals for removal and rapid biological decay of the dying material. Most grasses have growth points close to the ground, out of harm from grazing animals, so the animals could

remove leaves without damage to the plant and also allow sunlight to reach the growth points the following season. But if there are not enough large herbivores with vast microorganism populations in their moist gut to cycle this mass of material, then it remains standing.

As dead grass parts stand, without large herbivores to bring about decay, a chemical process of oxidation takes over. Oxidation is a very slow process: experimental plots show that it can take over 60 years for a perennial grass to finally break down if totally protected from fire or animals. Slow oxidation/weathering leads to premature death of most animal-dependent perennial grass plants, and thus to loss of biomass and diversity. Healthy grassland gives way to woody tap-rooted plants that are not dependent on grazing herbivores (if rainfall is high enough). And in places where rainfall is low, it gives way to desert bushes and bare soil—i.e., desertification.[121]

As discussed earlier, grazing animals are absolutely essential to properly functioning ecosystems. A March 2014 grazing impacts study by the University of Maryland provides just one example of research that has proven it. Researchers carried out field tests at sites on six continents, creating various adjacent test plots that either allowed or excluded grazing to specifically catalog grazing's effects. They consistently found greater plant diversity flourishing where animals were allowed to graze compared with plots where animals were excluded. The scientists concluded that because animal grazing allows more sunlight to reach the ground, more diverse types of plants were able to grow.[122] Allan Savory would probably add that "animal impact" is at least as important.

In the earth's recent history, the age-old co-evolutionary cycle of animals and grasslands was severely ruptured. By about 10,000 years ago, human hunting had decimated native roaming animal herds over much of the globe. Mere remnants of those once mighty animal mobs linger today. Wild grazing herds are sparse, and very few of the world's large predators remain. These anemic populations cannot and do not have a similar impact on grasslands as the thundering herds of yesteryear. It is thus the *absence* of animals rather than their *presence* that has caused land to become increasingly arid and desertified in recent millennia. This is why Savory talks about the greatest damage coming from "over-resting."

Savory further argues the tools to affect large tracts of land, such as the world's grasslands, are limited. Humans have at their disposal the following: technology (such as vehicles and other machines, and chemicals); resting (which can occur as part of a grazing regime); fire; and grazing.[123]

Of these alternatives, Savory believes cattle offer the best hope for reestablishing the necessary balance. Managed correctly, cattle herds carry out many of the ecosystem services once performed by wild herds. "We've got to *use* animals—their mouths to graze, their hooves to lay litter, to cover soil, to chip the surfaces, and more than anything else we're using the micro-organisms that are in their gut," Savory explains. "We are using the livestock as a proxy for the wild animals that were there before . . . Nothing else in the world is available to do this except livestock."[124] He urges, "There is no other known tool available to humans with which to address desertification."[125]

What's particularly intriguing about Savory's ardent advocacy for cattle grazing is that it comes from an ecologist, not a rancher. He has, he says, never been passionate about livestock. Instead, he has a lifelong, unwavering dedication to wildlife. Only after years of working to restore game preserves did he realize that to protect wildlife, livestock were essential.[126]

Savory now lives mostly in the United States, but he remains strongly connected to his roots in Africa. In those parts of the world where the desertification crisis is acute, Savory notes, expanding deserts are not only a climate change concern but also a major cause of conflict, poverty, and emigration.[127] At the Africa Centre for Holistic Management in Victoria Falls, Zimbabwe, a 200-hectare "learning site," Savory's methods are on full display. There, managing for benefits to wildlife, not livestock, they have increased the number of cattle year after year.[128] Cattle are indispensable, according to Savory. "We *need* cattle on the lands to stop and reverse desertification," he says.[129] "Leaving it to nature doesn't work because [the ecosystems] aren't natural anymore."[130]

Savory's global following is growing because he has demonstrated that what he is advocating works, and the results have been nothing short of astounding. On tracts his organizations directly manage, as well as on the millions of acres where his teachings have been put into practice, land that had dried out and begun turning to desert has been restored to grasslands teeming with water and life. Although his climate change arguments run counter to mainstream thought, they are simply too compelling to be ignored.

Several empirical studies have documented multiple ecological benefits to using methods like those advocated by Savory. In one such study, researchers at Texas A&M University, led by world-renowned range ecologist Dr. Richard Teague, compared the effects of different grazing methods, including a multi-paddock grazing with high stocking density (*mob grazing*), on several ranches that had each used the same management methods for at least nine years. The research team found wide-ranging benefits to the mob grazing method, including the following: less bare ground; more soil aggregation; soil that is more penetrable; lower sediment loss; higher soil organic matter; and the highest fungal/bacterial ratio, "indicating superior water-holding capacity and nutrient availability and retention." Overall, the researchers concluded: "This study documents the positive results for long-term maintenance of resources and economic viability by ranchers who use adaptive management and [mob] grazing . . ."[131]

Similarly, a multiyear field study done by researchers at Idaho State University compared different land management strategies for their effects on soil-water content. The three strategies were simulated Holistic Planned Grazing (the Savory approach), rest-rotation (lighter stock density grazing), and total rest (no grazing). Soil-water content was measured continuously for two years using 36 sensors. The researchers found both that the percentage of ground covered by plant litter was the highest and that the soil-water content was highest using the Savory method. "Management decisions (grazing and rest) can have substantial influence upon soil-water content and . . . soil-water content can vary substantially as a result of animal impact and the duration of grazing," the study concluded.[132]

Probably the most rigorous empirical study of cattle grazing's carbon sequestration and life-cycle impact was completed in 2018 by researchers at Michigan State University, which has a long and storied history of agricultural research. The team's field measurements found a drawdown of 1.5 tons of carbon per acre per year (1.5 tC/ac/yr) from cattle on grass, which is a lot—actually enough sequestered carbon to fully offset beef's greenhouse gas (GHG) emissions. The researchers note that enteric CH_4 emissions from grazing systems have garnered unfavorable attention. But, they point out, measurements of soil organic carbon have rarely been part of the accounting. "In fact," the Michigan State team says, "if soil C is sequestered through best management practices, our results suggest that enteric CH_4 from the finishing phase can be substantially mitigated . . . [O]ur results show that not only can adoption of improved grazing management facilitate soil C sequestration, but that the finishing phase of the beef production system may serve as an overall GHG sink." Carefully managed grass-based beef finishing can deliver environmental benefits (such as soil C sequestration and other ecosystem services), the study concludes: "if effectively implemented over a large area, total C sequestration in the Upper Midwest could increase substantially."[133]

In 2019, another important analysis of beef cattle's carbon footprint was released. It found cattle at White Oak Pastures farm to sequester more than enough carbon to offset *all emissions* in their beef production chain. They are, in other words, making carbon-neutral beef. This happens to be a beautiful, vibrant farm I know and love, in Bluffton, Georgia. Will Harris and his family keep adding new animals and crops, innovating with new products, and employing more people in the community. The farm raises vegetables, cattle, sheep, chickens, turkeys, ducks—to name just some of what they do. Animals move across fields as part of diverse rotations that resemble the complexity of nature.

White Oak hired the well-known environmental research firm Quantis to conduct a life-cycle assessment (LCA) of its farming operations. Quantis was asked to "account for the energy and environmental impacts of all stages of a product's life cycle, such as [the] acquisition of raw materials, the production process, handling of waste byproducts, and more." Of all its diverse crops and livestock, beef fared the best. "White Oak Pastures offsets at least 100% of our grassfed beef carbon emissions and as much as 85% of the farm's total carbon emissions," the farm announced in its blog. "The LCA analyzed the greenhouse gas footprint of our farm, which included enteric emissions (belches and gas) from cattle, manure emissions, farm activities, slaughter and transport, and carbon sequestration through soil and plant matter." White Oak Pastures attributed this success to its adoption of holistic management practices. "Our grassfed cattle sequester more carbon than they produce," the farm proudly announced when the environmental audit was completed.[134]

Another extraordinary paper, the most downloaded manuscript in the *Journal of Soil and Water Conservation* for 2016, argues that converting cropland in the Upper Midwest to well-managed grazing could lead to substantial climate benefits. A collection of rangeland experts, sustainable agriculture experts, and soil scientists collaborated on the paper, which showed that good grazing builds soil carbon, removes substantial carbon from the atmosphere, and is better for the climate than crop production. Specifically, they estimate a drawdown of 1.2 tons of carbon per acre per year (1.2 tC/ac/yr) via properly managed grazing, and that the drawdown potential of North American rangelands and pasturelands is 800 million tons (megatonnes) of carbon per year (800 MtC/yr). The authors argue that if crop production were replaced with well-managed grazing, the greenhouse gas emissions of agriculture would actually decline. And they estimate that if conservation approaches were completed on 25 percent of US crop and grasslands, the entire carbon footprint of North American agriculture could potentially be mitigated.[135]

Now the burning question many have been asking is whether the potential impact upon climate change of Savory-type methods, if broadly adopted, can be quantified. Fortunately, credible efforts have been underway to do just that.

One is a report by Seth Itzkan, an engineer who heads a consulting firm called Planet-TECH Associates. The firm's work includes land restoration and climate change mitigation. Together with his friend, Karl Thidemann, he co-founded the nonprofit group Soil4Climate, a Vermont-based organization focused on getting carbon back into soils as a key solution to climate change. I have worked with Seth and Karl for several years and value them both for their work and as friends.

Several years ago, Seth became so intrigued by the potential for grazing as a way to fight climate change he traveled to the holistic management center in Zimbabwe, staying for six weeks. He then returned for two more visits, each time for four to six weeks, to study Savory's planned grazing methods and their impacts on land, water, and wildlife. Combining observations and field research in Zimbabwe with extant research on soil carbon capture, Itzkan published a 34-page technical paper in April 2014 calculating the effect Savory's grazing methods could have globally for ameliorating climate change.[136]

Seth Itzkan's report concludes that the potential for carbon capture of Savory's grazing strategies is enormous. He calculates holistic grazing can sequester between 25 and 60 tons of carbon per hectare per year in semi-arid grasslands. As more and more carbon is captured, land would shift from semi-desert shrubland to healthy tropical savanna and perennial grassland. Itzkan also calculates that the total potential for capturing soil organic carbon in grasslands is around 88 to 210 gigatons, which translates to approximately 41 to 99 ppm atmospheric CO_2. In other words, widespread adoption of Savory's methods could sequester enough carbon in the soils to reduce CO_2 atmospheric load by nearly a third. This, the author notes, is "enough to dramatically mitigate global warming."

Cattle and Climate: The Final Word

Vibrant soil biology is the chief cornerstone of regenerative farming. A robust and growing body of research shows that cattle impact helps sequester carbon and build healthy, biologically active soils. In this way, well-managed cattle are reversing carbon's migration from soils to atmosphere.

It's easy to feel discouraged by the prevalence of reductionist thinking in public discourse. Beef still gets pilloried on a daily basis for its purported role in climate change. In the past few years, however, the rising chorus of thoughtful, intelligent voices who are objecting has been heartening.

One was a letter to the editor in *Grist* from a world-famous vegetable farmer, Eliot Coleman, a former vegetarian. Responding to a *Grist* article suggesting meat bears a disproportionate share of the blame for global warming, Coleman wrote:

> *When I think about the challenge of feeding northern New England, where I live, from our own resources, I cannot imagine being able to do that successfully without ruminant livestock able to convert the pasture grasses into food. It would not be either easy or wise to grow arable crops on the stony and/or hilly land that has served us for so long as productive pasture. By comparison with my grass fed steer, the*

soybeans cultivated for a vegetarian's dinner, if done with motorized equipment, are responsible for increased CO_2 . . . Targeting livestock as a smoke screen in the climate change controversy is a very mistaken path to take since it results in hiding our inability to deal with the real causes [burning of fossil fuels].[137]

Like other agricultural practitioners I've met, Coleman well understands that growing food is messy and complex, much more so than many non-farmers care to consider. He appreciates, too, the unique and beneficial role of ruminants, especially if farming is to be done in a truly regenerative way. I particularly appreciate Coleman's message that focusing on cattle emissions is a red herring that takes attention away from where it should really be: fossil fuels.

To date, efforts to accurately account for cattle's greenhouse gas emissions have fallen short. The most notable attempt, *Livestock's Long Shadow*, presented bloated figures. Including deforestation from developing countries as part of total cattle emissions is unfair and unreasonable. In any case, the report's numbers, especially for cattle, should be taken with a giant grain of salt because they utterly failed to account for carbon sequestration. This is a serious flaw because cattle grazing has the potential to significantly or completely offset all beef production emissions.

With all of that said, FAO is the most credible entity doing such calculations. So I use its numbers to roughly calculate the current global contribution of all cattle at 8.5 percent. Using FAO's 14 percent number for emissions from all animal-based foods, and using a "CO_2 equivalence" for beef and milk production from cattle respectively at 41 and 20 percent of all livestock sector emissions (also from the report),[138] means that foods from cattle are responsible for 61 percent of the 14 percent, or in other words 8.5 percent of greenhouse gases. Although these may be roughly the current values, that's not to suggest these emissions are *necessary* to milk and meat production.

At various points, FAO's report also shows that emissions from cattle are not unavoidable. As it notes, "emissions from energy consumption on farms and in processing are negligible in beef and limited in dairy" (about 8 percent of the sector's entire emissions).[139] It attributes the single largest portion of emissions to animal feeds.

"Feed production, processing, and feed makes up the entire sector's biggest share, at 45 percent."[140] Yet the special gift of cattle is that they don't actually need to be fed. Like wild ruminants, they can survive entirely on grasses and other naturally occurring vegetation. In other words, feed emissions can be reduced or, eventually, even eliminated. The large share of emissions from livestock feeds at once reminds us of the special value of grazing animals and reinforces the urgency of getting grazing animals back onto grass.

FAO itself acknowledges beef and dairy emissions can be offset. Getting cattle onto grass is part of the solution. Improving grazing practices is another part. Carbon sequestration in soils triggered by animal impact can substantially counterbalance the sector's emissions. When offsets are considered, greenhouse gas emissions from cattle have the potential to be minimal, zero, or even a negative number. Whatever the correct number, cattle's impact on climate must always be considered in the greater context of the tremendous nutritional, ecological, and social benefits of having cattle in our food system.

Moreover, it really only makes sense to talk about emissions of specific foods if we consider the greater context of what food is really all about. And that's nutrition. Precisely this type of analysis has been carried out by Dr. Michael Lee, professor of sustainable livestock systems at Bristol University and head of sustainable agriculture sciences at England's Rothamsted agricultural research center. Dr. Lee, whose particular expertise is grass-based livestock farming, emphasizes, "Sustainability is complex." But he notes that the standard metric for measuring a food's global warming impact (like that used by FAO) is simplistic: just "CO_2 equivalents per kilogram of product." This, Lee argues, is inherently flawed. "We don't consume kilograms of product, we consume nutrients." He urges consideration of several additional factors, starting with a food's nutritional value. Among other things, he considers protein, omega-3, and micronutrients iron, zinc, calcium, selenium, and B vitamins. Beef is especially nutrient-dense. So when recommended daily intake (RDI) is taken into account as part of kilograms per CO_2 equivalent, the results shift considerably. The conventional accounting system suggests that factory-farmed chicken and pork are more "climate-friendly" than grazing animals. But when nutritional value is included, everything changes. Beef, Dr. Lee has demonstrated—especially grass-based beef—performs the best.[141]

Because well-managed cattle benefit soils, analysis of their global warming impact should also include consideration of that effect. The writings of soil scholars Lal and Stewart point to two possible approaches for climate change mitigation. First, limit greenhouse gas emissions. Second, improve the uptake of greenhouse gases from the atmosphere into stable pools.[142] To stop the catastrophic effects of global warming, we surely must pursue both. As this chapter shows, soils hold enormous potential for the second part of the approach. And because cattle impact boosts soil's carbon transactions, cattle are essential to the strategy.

Beyond climate change, Lal and Stewart also note that improving the world's soils is "an important and necessary step to free much of humanity from perpetual poverty, malnutrition, hunger, and substandard living." The key, they believe, is focusing on soil's physical and biological fertility. "Conservation agriculture that combines reduced tillage, crop residue retention, and functional crop rotations, together with adequate crop and system management, permit the adequate productivity, stability, and sustainability of agriculture," they say.[143] Chemical fertilizers, in other words, will not do. In May 2014, when queried by *The Boston Globe*, Lal said he considers Allan Savory's methods, using carefully planned mob grazing, an effective way to increase soil carbon content.[144]

What's certain is that efforts to minimize greenhouse gases must be far more sophisticated than blanket condemnations of beef. Research clearly shows that varying methods of raising animals lead to differing global warming impacts.

This "It's not the *cow*, it's the *how*" variability is, in fact, explicitly acknowledged in FAO's most recent climate change report, which states:

> *For ruminant products especially, but also for pork and chicken meat and eggs, emission intensities vary greatly among producers. Different agro-ecological conditions, farming practices and supply chain management explain this heterogeneity, observed both within and across production systems. It is within this variability—or gap between producers with highest emission intensity and those with lowest emission intensity—that many mitigation options can be found.*[145]

Variability in climate impact is true for every food we eat, not just for meat. A study by Sweden's national environmental agency showed that, depending on how and where a vegetable is produced, its carbon dioxide emissions vary by a factor of 10.[146]

It should also be kept in mind that what happens on farms is only part of any food's greenhouse gas emissions. Only about one-fifth of the food system's energy use is farm-related, according to University of Wisconsin research.[147] In the U.K., the Soil Association estimates that only half of food's total greenhouse impact has any connection to farms.[148] The rest comes from processing, transportation, storage, retailing, and food preparation.[149] The seemingly innocent potato chip, for instance, turns out to be a dreadfully climate-hostile food, mostly because of what's done to the potato after it leaves the farm—it's transported, stored, cooked at high temperatures, packaged, and transported again, all for very little nutrition.[150] Foods that are minimally processed, in season, and locally grown, like those available at farmers markets, farm stands, and backyard gardens, are generally the most nutritious and climate-friendly.

Rampant waste at the processing, retail, and household stages further compounds the problem. Almost half of the food produced in the United States is thrown away, according to University of Arizona research.[151] The single largest share of US household landfill waste is food. This wasted food is a double whammy against the environment. Each step of creating it—growing, transporting, storing, and processing—causes emissions. Then the food item goes on to release methane as it decays in a landfill—*all without ever having nourished anyone*! Thus, each of us can also reduce our personal global warming impact simply by more judicious grocery purchasing and use.

The reality is that none of us—from vegan to carnivore—can avoid eating foods that play a role in global warming. Because the effects of any given production system are so variable, and because cattle grazing offers enormous potential for climate change mitigation, it is both unhelpful and misleading to single out beef.

Instead, all consumers, whatever the dietary preferences, should be encouraged to eat in ways that support both their own health and good farming practices. Our purchases can support farming that fosters building soils and sequestering carbon. We can compost our food wastes.

We can grow a portion of our own food. When Americans were urged to plant "Victory Gardens" during World War II, they collectively grew about 40 percent of their vegetables.[152] We should especially eat less industrially produced food and eat *better* meat.

Even though I believe in the value of making better food choices, I want to take a moment to stress that it will never be enough to effectively alter the course of climate change. We make an important personal decision when we grow our own food, and when we buy from good farms. What kind of food we eat and where it comes from affect our personal health; they can ameliorate our peace of mind. But focusing on our personal global warming contributions will fall far short of mitigating the major sources of greenhouse gas emissions. Our diets have far, far less impact on the climate than does US policy. This has been made clear by climate commentators like by David Wallace-Wells in his book *The Uninhabitable Earth: Life After Warming*. He argues the climate crisis is urgent and imminent. We need to mobilize all tools at our disposal to stop climate change, Wallace-Wells urges; strategies like carbon taxes, carbon capture, and green energy will have far greater effect than any choices we make as consumers.[153]

Although it has far less real-world impact, I agree wholeheartedly with seeking to reduce our individual contributions to global warming. And I certainly strive to do it myself. To that end, the most important thing we can do is cut back fossil fuel consumption. We can fly less, drive less, buy electric vehicles, ride our bicycles. We can reduce our overall consumption of goods. Clothing, for one. In 2019, the World Bank warned, "The fashion industry is responsible for 10% of annual global carbon emissions, more than all international flights and maritime shipping combined. At this pace, the fashion industry's greenhouse gas emissions will surge more than 50% by 2030."[154]

As part of our personal actions, we can also reduce our dietary carbon footprint. Ways to do this include avoiding processed foods and those from industrialized operations; reducing our food waste; growing as much as we can; and buying local and in season. Each of these changes would also make our diets more healthful, so it's a sensible thing to do on several levels.

Regardless of what we eat, though, it is absurd to focus narrowly on a food's climate impact to the exclusion of all else. As this entire discussion has hopefully made clear, every part of the agricultural system—whether raising carrots, soy, or beef—has many and varied effects on humans, animals, and the environment. I wholeheartedly and enthusiastically embrace the idea that food choices matter. As much as we are able, we should each choose wisely every day, with consideration not just of taste but also of health and of the many ripple effects of the systems we are supporting. However, I soundly reject the suggestion that our ethical obligations start and end with climate change impacts.

Climate change is important and urgent, yes, but it is only one of many vital issues related to the health of our bodies and the planet. Other parts of this book are dedicated to examining issues beyond climate change. For each of us individually as well as collectively, climate change effects should be one of myriad factors we consider. As a society, we must plan food systems that create nutrient-rich, health-supporting foods while being ecologically sound and regenerative, as well as humane to workers and animals. As eaters, we owe it to ourselves to seek foods that are delicious, healthy, and ethically produced.

As with the production of all foods (and all consumer goods), raising cattle and turning them into beef has many and varied impacts—both positive and negative. The only sensible way to evaluate beef is through consideration of these myriad effects.

Finally, if you wish to help shape the food system for the better with your food choices, quitting beef would have far less impact than shifting from commercial beef to well-raised beef. The global market for commodity beef is massive. The purchasing dollars of any of us are like a drop of water in the ocean. But when you buy grass-fed beef from a well-run farm or ranch in your region, you may *literally* be providing the financial support that enables survival for that family business. Rebuilding our food system will happen one farm at a time.

How, then, should we think of beef's role in climate change? First, we should always remember that cattle are living organisms intimately connected with their ecosystems. When we equate animals with machines, we quickly lose our way. Cows are not cars. Nature is a dynamic system in which plants, animals, and fungi are all connected. It is not only a refuge where we can go for calming and inspiration, but also a boundless font of knowledge. "Look deep, deep into nature, and then you will understand everything better," Albert Einstein said in 1951, near the end of his life.[155]

Cattle are domesticated, yes. But they are still animals in ecosystems. They are part of nature. Physicist and philosopher Fritjof Capra has said, "Everything in nature is connected. Only human-made machines are linear." The issue is far more complex than bumper stickers or FAO press releases would have us believe.

Early-20th-century journalist H. L. Mencken famously said, "For every complex problem there is an answer that is clear, simple, and *wrong*." We need to stop viewing cattle through the narrow lens of a climate problem like fossil fuel emissions that needs elimination or even mitigation. Instead, we should focus our energies on fully tapping into their unique superpowers: Domesticated ruminants can heal our damaged environment. They can draw carbon from the atmosphere back into soils, all while generating wholesome, nutrient-rich foods. If we want a slogan, perhaps it should be: *Beef: Bringing Carbon Back to Earth.*

CHAPTER 2

All Food Is Grass

MODERN URBAN DWELLERS TEND TO LOOK AT GRASS AND SEE ONLY monotonous, ubiquitous, inedible green ground cover. Environmentally minded landscapers loathe it and have taught us to be skeptical of expansive, watered, suburban lawns. These days few of us regard grass as part of our food system, let alone its foundation. Grass's connection to what we eat seems remote, if we think of it at all.

Why should a book on beef dwell on grass? Evidently, cattle eat it. But that's just the beginning of the story. In the previous chapter, we explored how leaves and roots of grass, working with fungi, glomalin, and soil microorganisms, as well as whole grassland ecosystems, all play critical roles in staving off our planet's warming. And despite our general lack of appreciation, grasses are the most important plants in the world. They are also inextricably linked to cattle.

Grass covers around 40 percent of the earth's land surface, about 70 percent of the world's agricultural area.[1] It is the fourth-largest plant family in the world, with more than 11,000 species worldwide.[2] But grass is mostly cellulose—poor in nutrients and nearly indigestible.

The bovine, however, has a special gift: a rumen that allows it to live off little more than grass. Cattle have a unique ability (you might even say superpower) to survive on this widely available, naturally occurring vegetation without being provided feeds.

This special trait has made cattle essential to humans in many climates and geographies for thousands of years. They have accompanied human migrations; they have fertilized our fields and gardens, powered our plows and carts, covered our feet and clothed our bodies with their hides, provided milk and meat—all while living primarily or entirely on a simple diet of grass.

Cornell University professor of agriculture and ecology David Pimentel, who died in 2019, was famously critical of the food industry, particularly the meat sector. Yet his classic treatise *Food, Energy, and Society* (co-authored with his wife, Marcia) repeatedly emphasizes the irreplaceable role of cattle and other grazing animals for human survival. "Cattle, sheep, and goats will continue to be of value," they noted, "because they convert grasses and shrubs on pastures and rangeland into food suitable for humans. Without livestock, humans cannot make use of this type of vegetation on marginal lands."[3]

Grass is, in fact, the base layer of the global food system. It steadily performs the job of converting massive quantities of solar energy into food usable by grazing animals and, eventually, humans. Grasses and herbivores, functioning together, are the indispensable intermediaries between humans and the energy of the sun.

Their invaluable cohorts, of course, are water and soil. Earth and space sciences professor David Montgomery admiringly describes soil as "the dynamic interface between geology and biology, the bridge between the dead world of rock and the bustling realm of life."[4] He also calls soil "the thin layer of weathered rock, dead plants and animals, fungi, and microorganisms blanketing the planet," which "has been and always will be the mother of all terrestrial life." Dr. Montgomery considers it "every nation's most critical resource, one that is either renewable or not, depending on how it is used."[5]

Soil, Montgomery says, connects "the rock that makes up our planet and the plants and animals that live off sunlight and nutrients leached out of rocks," which, "in the big picture . . . regulates the transfer of elements from inside the earth to the surrounding atmosphere."[6] It provides grass with essential nutrients while acting as the medium by which oxygen and water are retained and supplied. "Acting like a catalyst, good dirt allows plants to capture sunlight and convert solar energy and carbon dioxide into the carbohydrates that power terrestrial life right on up the food chain."[7]

From elementary school through college, I was often taught about an ongoing competition for survival in the wild. But ancient wisdom and modern research reveal natural systems to be at least as much about cooperation. Everything in nature is connected. Everything works in systems.

Together soil and water enable grass to grow, while grass, in return, performs essential services for them. On average, 90 percent of grass's bulk is belowground, in the form of long, filamentous roots—a tangled network of strands tenaciously gripping the soil. "Grass prevents erosion by binding to the soil," explains the geology textbook *People and the Earth*. The hold of grass roots on soil prevents it from being washed or blown away. Through complex biological processes, grass assists the formation of new soils. By creating a vast network of tiny underground channels, and by making soil aggregation possible, the grasses of the world also render the earth a far more effective keeper of water.

When grass is plowed under, however, this gripping tapestry of filaments is torn to shreds and the intricate network of water channels vanishes. "Replacement of grassland with cropland in the American Midwest, its 'breadbasket,'" for example, has "caused nearly one-third of the topsoil to be stripped by erosion in the past 100 years."[8]

Even a casual observer can see that grasses feed wild and domestic herbivores. It may be less obvious that the relationship is two-way: As herbivores are nourished by grass, their grazing and trampling maintain and even regenerate grasslands. Pruning by animal mouths supports plant growth just as it does in your garden. Grazing and trampling also suppress the emergence of woody plants that transform grasslands into ecosystems less hospitable to grass and grassland creatures. When grazing animals are extirpated or otherwise disappear from a terrain, the ecosystem, including all life that depends on it, is radically altered. "When grazers are removed, grasses lose their competitive advantage and forbs and shrubs quickly become established," explains environmental resource management professor J. P. Curry.[9]

Just as mowing helps keep a lawn grassy, lush, and densely vegetated, so animal grazing triggers plant growth by pruning off old and dead parts and by exposing plants' growth points to more sunlight. (If you've ever left a portion of your lawn unmowed for a while, you've witnessed this effect firsthand.)

But grazing animals have a far more potent effect on plants and soils than a lawn mower ever could. Their hooves help break up compacted soils and push seeds into the ground, preventing the seeds from getting blown or washed away, or gobbled up by hungry birds or field mice.

By pressing plant leaves and stems into soils, their hooves also catalyze cycles of decay. Down in the dirt, vegetation becomes enveloped by and feeds the decomposing microbes that power carbon and nutrient cycling.

Cattle also outperform mechanical mowers because they ensure the continual addition to the soil of moist, nutrient-rich organic matter, long regarded by farmers as agricultural gold—known to many non-farmers as cow patties. "Increasing soil organic matter by applying manure or similar materials can improve the water infiltration rate by as much as 150%."[10]

Cycling vegetation through the bovine's unique, multi-chambered stomach hastens a grassland's cycle of biological decay, which boosts the vitality of the entire ecosystem. Topsoil's billions of microorganisms help plants get nutrients from organic matter and mineral soil. Soil organisms "supply plants with nutrients by accelerating rock weathering and the decomposition of organic matter."[11]

Grazing thus bolsters continuous regeneration of soils and supports grass's diverse functions, from the microscopic to the ecosystem scale. Dense mats of sod shield soils from being blown away by wind and being washed away by water. Tangled grass root masses hold soils in place, improving their ability to hold moisture.

"There are no better soil stabilizers than luxuriant pasture grasses and legumes, range grasses, and forest trees and shrubs," states university textbook *Soil and Water Conservation*.[12] First, because aboveground plants protect soils against the force of falling precipitation and sheeting water. "Close-growing perennial vegetation intercepts rain drops, thus reducing impact energy and decreasing soil dispersion and splash and sheet erosion. The velocity of surface water flow is reduced by contact with plant stems and residues." Second, because water soaks into the ground where grasses are growing, and soil particles are caught by grass blades rather than running off. "The result is more and faster infiltration of water into deep soil horizons, clean water flowing slowly along the soil surface, and less soil erosion and sediment." Finally, plant roots cling to soil, protecting it from wind. "Vegetation also stabilizes soil against wind erosion by the 'holding' action of the plants roots and by decreasing wind velocity at the soil-atmosphere interface."[13]

A review of studies of world soils reported field trials that found nearly complete absorption of water (what Savory would call effective rain) with dense vegetative cover. By contrast, only 20 percent of rain infiltrated in bare dirt.[14]

Even in dormant seasons, grasses safeguard soils. "Land areas covered by plant biomass, living or dead, are more protected and experience relatively little soil erosion because raindrop and wind energy are dissipated by the biomass layer and the topsoil is held by the biomass," according to the Pimentels.[15]

Wendell Berry eloquently contrasts a crop field and prairie this way:

The most noticeable difference is that whereas the soil is washing away in the cornfield, it is building in the prairie. And there is another difference that explains that one: the corn is an annual, the cornfield is an annual monoculture, but the dominant feature of the native prairie sod is that it is composed of a balanced diversity of perennials: grasses, legumes, sunflowers, etc., etc. The prairie is self-renewing; it accumulates ecological capital; and by its own abounding fertility and diversity it controls pests and diseases. The agribusiness cornfield, on the other hand, is self-destructive; it consumes more ecological capital than it produces; and because it is monoculture, it invites pest and diseases.[16]

Farm and soil experts agree: The best way to protect soils and water is continuous plant cover—permanent pastures and rangelands. The Pimentels note: "Vegetative cover is the principal way to protect soil and water resources."[17] Wendell Berry writes: "If you want to stop soil erosion (so the prairie shows) you have to keep the ground covered—all the time, winter and summer. If you want to keep the ground covered all the time, the best way to do it is with a diverse, mutually beneficial polyculture of perennial plants."[18]

Compared with such grass-covered areas, "cropland is more susceptible to erosion because of frequent cultivation of the soils and the vegetation is often removed before crops are planted,"[19] *Food, Energy, and Society* notes. Fields used for growing crops are much more exposed to wind and rain energy because they are often partly or totally bare. Between plantings, the fields are naked, vulnerable to the forces of wind and rain. Once crops have been planted, seeds and seedlings offer little protection. Even when crops are mature, "[r]ow crops are highly susceptible to erosion because the vegetation does not cover the entire soil surface."[20] On average, according to *Food, Energy, and Society*, erosion rates on US pastures are 40 percent lower than on US cropland.[21]

Other studies have found still more dramatic differences. Environmental nonprofit Land Stewardship Project field trials showed that compared with cropland, perennial pastures used for grazing can decrease soil erosion by over 80 percent.[22] Soil experts Lal and Stewart state: "Erosion rates from conventionally tilled agricultural fields average 1 to 2 orders of magnitude greater than erosion under native vegetation, and long-term geological erosion exceeds soil production."[23]

Even as part of a crop rotation, grasses, especially when grazed, provide manifold benefits. By delivering organic matter and carbon, grass plants feed the billions of tiny organisms teeming below the earth's surface. These microorganisms increase soil fertility by making nutrients bio-available for plants. Grazing, by stimulating plant growth and by cycling plant matter through ruminant guts, assists such carbon–nutrient exchanges. When plants are better nourished, and soil microbes are diverse and abundant, plants are more resilient against diseases, pests, and harmful fungi. Before there were agricultural chemicals—fertilizers, pesticides, and fungicides—there was grass.

Adding manure to the land enhances soil's chemical properties in several ways, according to soil gurus Lal and Stewart. First, manure directly provides carbon and nutrients, including nitrogen, phosphorus, and potassium. Second, manure improves soil's cation exchange capacity—its relative ability to store nutrients.[24] And manure stimulates beneficial biological activity. As noted earlier, chemical fertilizers simply cannot provide comparable benefits. "Mineral fertilization without organic amendments leads to the mineralization of [soil organic matter] and to a decrease in soil structure, pH, and the attendant decline in agronomic productivity," Lal and Stewart report.[25]

The generous gifts of grass make it the stalwart of regenerative food production. Grassy areas with properly managed cattle herds offer a stunning opportunity to sequester carbon, reverse desertification, foster all life in an ecosystem, and mitigate climate change. Ideally, then, all food is grown with grass.

Earlier peoples did not know the details of complex molecular-level interactions. Yet their traditions cherished soils and grass as sustenance for the animals on which they depended. "Out of earth you were taken, from soil you are and unto soil you shall return," says the book of Genesis.[26] The Psalms praise God in saying: "You make grass grow for flocks and herds and plants to serve mankind; that they may bring forth food from the earth."[27] In Deuteronomy, God tells His people: "And I will send grass in the fields for thy cattle that thou mayest eat and be full."[28]

In 1872, Senator John James Ingalls of Kansas delivered a speech called "In Praise of Bluegrass," in which he poetically echoed such biblical themes. "The primary form of food is grass," said Ingalls. "Grass feeds the ox: the ox nourishes man: man dies and goes to grass again; and so the tide of life, with everlasting repetition, in continuous circles, moves endlessly on and upward, and in more senses than one, all flesh is grass."[29] Just a few years earlier, the great German composer Johannes Brahms completed his A German Requiem based on the same biblical passage, from the First Epistle of Peter: "For all flesh, it is as grass."

To fully appreciate how intimately connected we are with grass, we must look even much further back in time, before the appearance of *Homo sapiens*. Prehistoric earth was not, as is commonly believed, thickly carpeted from ocean to ocean with damp ferns and towering trees. Ancient flora was highly varied, and from at least 65 million years ago, it included grasses.[30] Probably not coincidentally, and likely due to climatic changes, grasses emerged around the time of the disappearance of dinosaurs, which had previously populated the earth for 160 million years.

From approximately 20 to 10 million years ago, grasses proliferated in earnest. Changes in the earth's climate are again likely responsible. Carbon dioxide levels dropped, followed by a period of time in which atmospheric carbon dioxide sharply increased, to about 400 ppm (somewhat higher than today's level).[31] Amid such fluctuations, grasses had a photosynthetic advantage over other plants. Around the same era, forest-clearing wildfires became common occurrences on the earth. Into these newly opened spaces, grasses (along with forbs, other non-woody vegetation) filled the voids. Resulting from some combination of atmospheric changes and forest fires (scholars debate which factor was more significant), grasses carpeted large swaths of the globe.[32]

The millennia that followed saw the co-evolution of grasses with a dazzling array of prehistoric animals, including grazing herbivores and the predators pursuing them. For some five million years enormous populations of "megafauna," great beasts that lived either entirely or mostly from grasses or ate animals that did, roamed much of the globe. This included lands that are now the United States.

As I gaze out at the open meadows of our California ranch, I sometimes picture the mind-bending prehistoric menagerie that once populated these lands. The western camel, with limbs 25 percent larger than modern dromedaries; the large-headed llama; elk, deer, and antelope; two varieties of bison (including the giant bison, the largest one that ever lived); the shrub ox and woodland musk ox (which, incidentally, both resemble modern cattle); the mountain goat; the bighorn sheep; two types of horses; and two varieties of mammoths.[33] Like modern elephants, mammoths abetted forest clearing by felling trees on which they were dining and by effectively crushing aspiring saplings with their weighty footsteps. Predators feasting on prehistoric grazing herds included a lion larger than those of today's Serengeti; a cheetah; a jaguar; a sabrecat (the size of a female African lion); a puma; wolves; and three types of bears.[34] Not only did these diverse animals occupy and rely upon grasslands, but their bodies' impacts kept the open areas in existence.

Likewise, early humans depended on and co-evolved with grass. By the time *Homo sapiens* emerged, some 200,000 years ago, grasses flourished in nearly every geography and every climate. Some directly nourished humans with their seeds, which humans gathered and ate. More important nutritionally, grasses fed the prey of the human hunter and enabled successful hunting; in open, grassy areas, people could spot, track for long distances, and ultimately kill foraging herbivores.

Ample scientific evidence supports that humans and grasses co-evolved. We influenced how and where grasses developed and grasslands were a major factor in human evolution. *Born to Run*, Christopher McDougall's fascinating bestselling book about long-distance running among traditional peoples, expounds on the idea. Known as the Running Man theory, it argues that the human capacity to run very long distances (not just 26 miles, but 50, 75, and more)—farther than nearly any other animal—was the secret to our ancestors' hunting successes. In other words, not speed or strength but *endurance* was our species' special talent. But hunting based on outlasting your prey works only if there are vast open ranges where prey can be spotted and followed even after an initial burst of sprinting to escape. Thus, the theory propounds, open, grassy areas were essential to humanity's success as a species.

Grasslands also provided a host of other benefits to early humans— ranging from the purely aesthetic to vital self-defense. These included flowers for decoration, fiber for basketry, and the unobstructed views that permitted humans living in settlements to see approaching danger from great distances.

These life-sustaining functions motivated humans throughout the world to both create and maintain open grassland areas. Long before settling and farming the Fertile Crescent some 12,000 years ago, humans had been actively managing their surroundings to support grass. Like people in the Americas, their management tools included clearing and burning.

Populations of the world's prehistoric mega-beasts precipitously declined around the dawn of agriculture. Nonetheless, these many years later they persist in shaping the world's plants, animals, and ecosystems. At Oxford University in March 2014, scientists from various disciplines gathered to focus on the unique ecological role of these large creatures. "[E]cologically speaking," the conference's co-organizer said, "it was only a blink of an eye ago that there were megafauna everywhere." Among scientists in attendance, there was a strong consensus that "megafaunal extinctions did have a huge impact on the structure of ecosystems, which still ramifies through to the ecosystems today." He added, "What we think [of] as natural now are still carrying many disequilibria—or ghosts—resulting from the loss of the megafauna in terms of their structure, functioning and nutrient recycling." Among other functions, researchers agreed these large animals would have distributed "vital nutrients for plants via their dung and bodies."[35]

The loss of these large herbivores and predators has weighed heavily on earth's ecosystems. So much so that credible US scientists, including some at Cornell University, actually advocate reintroducing large predators to the US plains. "Their disappearance left glaring gaps in the complex web of interactions, upon which a healthy ecosystem depends," they have stated.[36]

This is part of a recent branch of the environmental movement called *rewilding*, which has especially gained steam in Europe. Underlying the idea is the premise that ecosystems cannot properly function without large herbivores. A *New York Times* article described the movement as advocating for ecological restoration through the injection of wildlife, "particularly of large herbivores." Without big grazing animals, it states, "untended farmland will become overgrown with thick vegetation that will end up killing off what biodiversity still exists today." The article quotes the communications director for Rewilding Europe saying, "We need to bring in a few of the parts that are missing because they just aren't there anymore, and that missing part is often the large herbivores."[37] Among projects already under way are reintroduction of the European bison in an area of abandoned farmland in the Romanian Carpathian Mountains and reintroduction of the ibex along a stretch of the Adriatic coast of Croatia where there are two national parks.[38]

By about 100,000 years ago, many large animals had gone extinct, probably mostly from human hunting (although scientists disagree on the relative importance of various factors). In California, grassland experts believe that as recently as 6,000 years ago, 19 species of large browsing and grazing animals still populated the open lands.[39] Herbivores ate down vegetation while they, along with their predators, trampled the earth with their hooves and paws. Fires, both natural and human-generated, and the presence of these animals, established and maintained California's grasslands.[40]

Cattle, of course, are domesticated animals. They belong to the zoological order Artiodactyla (Greek for "even-toed"), suborder Ruminata (cud-chewing animals that are primarily herbivorous). Most modern cattle are considered descendants of a much larger, now extinct wild beast called the aurochs. The book *Animals That Changed the World: The Story of Domestication of Wild Animals* calls cattle domestication "the most important step ever taken by man in the exploitation of the animal world." But it notes that how and when it happened are still largely unsettled. Recent research suggests bovine domestication may have begun as early as 11,000 years ago, independently in two or more locations, probably India and the ancient Near East.[41]

Cattle first arrived in the Americas in 1493, with Christopher Columbus's second voyage, which brought Spanish breeds, mostly as draft animals. Not until 1591 did the original seed stock for Latin America's cattle herds arrive, when a Spanish merchant named Gregorio de Villalobos shipped a small group of Andalusian breed cattle to the New World. In the early 1600s, European settlers of North America began landing with early British and Continental breeds, primarily for labor and milk.

As settlers pushed south then pioneers headed west, cattle were part of the migration, proliferating particularly in areas with abundant native grasslands. Many early cattle ranches were in the South, especially Georgia and the Carolinas. "The long grazing season, mild winters, sparsely wooded uplands were especially favorable for beef production," notes a cattle history. "It was said that a steer could be raised as cheaply as a hen."[42]

By the mid-1800s, cattle ranches were being established on the grasslands of the West. But the bulk of the nation's cattle raising was taking place in the heart of the country. "Prior to the Civil War, the Ohio and upper Mississippi Valley states constituted the center of the beef cattle industry. On practically every farm of this area was a herd of beef [mother] cows."[43]

Europeans spreading across North America encountered thickly vegetated, biologically diverse grass prairies. Native plants acting in concert with huge herds of bison and other grazing animals—modern and prehistoric—had produced the most fertile and abundant topsoil on the planet. Settlers tore open these prairies with their plows. Cutting up the dense natural sod to plant crops was tantamount to exploding vaults holding our nation's most precious jewels. There is inestimable wealth caught up in native grassland soils—biological riches created from millennia of captured sunlight and animal impact. An old saying goes: "Breaking the sod will make a man, re-making it will break him." Plowing squanders a natural wealth that is nearly irretrievable.

As elsewhere in the world, poor farming practices eroded midwestern topsoil at an alarming rate. The first farmers to plow up these prairies had been greeted by chest-high grasses and topsoils six feet deep. By 1934, aggravated by years of serious drought and crop failure, a 100-million-acre dead zone on the same lands caused epic dust storms as far away as New York City. The American heartland became known as the Dust Bowl. In 1935, under Franklin Delano Roosevelt's leadership, the US government created the Soil Conservation Service in the hope of reversing this human-caused ecological disaster.

The new federal agency's field trials soon reinforced history's lessons on the perils of the plow. They demonstrated that breaking sod and stripping away the earth's protective vegetative cover disrupts soil's life-sustaining systems, depleting natural fertility, destroying the earth's ability to retain moisture, and killing beneficial organisms. The studies also showed the resulting brittle, lifeless soil was becoming less productive over time as it was suffering severely from wind and water erosion.

Mercifully, by the mid-20th century, Soil Conservation Service research was also uncovering solutions to agriculture's problems. The answers were found in the wisdom of natural systems. Most important, the agency's studies demonstrated nature regenerates itself with diversity. Regularly alternating croplands with pastures—especially of mixed grasses and clover—reverses much of the damage done by row crop cultivation. Such meadows feed grazing animals while also creating a permanent farming system that can produce food sustainably over the generations.

Cover crops were already well known at this time to benefit farms, especially by keeping soils in place. Blankets of aboveground leaves shield soil from the erosive forces of raindrops and winds. Simultaneously, tangled, fibrous networks of roots beneath the surface hold soil in place, feed soil organisms, and provide sponge-like channels for water.

The nitrogen cycle is an important part of this regenerative natural system. Clover and other legumes take nitrogen from the air. Assisted by microorganisms living in root nodules symbiotically with plant hosts, the legumes "fix" this nitrogen by reducing inert atmospheric nitrogen to biologically active ammonia. The legume-microorganism cohort then replenishes the soils with nitrogen in this usable form. "Once incorporated into soil organic matter, nitrogen can circulate from decaying things back into plants as soil microflora secrete enzymes that break down large organic polymers into soluble forms, such as amino acids, that plants can take up and reuse."[44] Most nitrogen enters soil from the atmosphere via this sort of biological fixation, according to Professor

David Montgomery. This vital function of soil microorganisms is in addition to their pivotal role in glomalin production.

These microscopic transactions had not yet been identified. Nonetheless, Soil Conservation research station investigations did show that mixes of grasses and legumes restored soil's fertility, organic matter, and tilth (soil's texture and ability to retain water), along with controlling insect pests and weeds. Their field trials found grass cover to be an impressive 200 to 2,000 times more effective than cultivated row crops in protecting earth from soil losses.[45]

Such research fostered optimism that US farming was at the dawn of a new era of economic and environmental sustainability. USDA's official *Yearbook of Agriculture* for 1948 was titled simply *Grass*. It hopefully described a newly emerging "grassland philosophy," touting the plant's forthcoming Golden Age.

The volume's foreword by Clinton P. Anderson, then secretary of agriculture, calls grass "our alliance with nature" and the "foundation of security in agriculture." He notes farming based on grass would improve national health and is "a tool against floods and a guardian of the water supplies." Anderson writes:

> *[M]any of the people with whom I have talked look upon grassland as the foundation of security in agriculture. They believe in grass, and so do I, in the way we believe in the practice of conservation, or in good farming, or prosperity, or co-operation. For grass is all those things; it is not just a crop. Grassland agriculture is a good way to farm and to live, the best way I know of to use and improve soil, the very thing on which our life and civilization rest.[46]*

Around the same time, the American Society of Agronomy's former president H. D. Hughes authored a hefty tome called *Forages*. In it he wrote, "That adapted grasses and legumes are the chief tools in soil building, improvement and conservation is now generally recognized."[47]

Looking at today's industrialized food production, these statements are nothing short of astonishing. They are remarkable both for their prescience of modern soil science and for their misplaced optimism. Soon after these words were written, historical events took over. American agriculture veered sharply in an altogether different direction. At the close of World War II, the United States had an abundance of munitions plants. Rather than being shuttered, they were converted to manufacturing agricultural chemicals. Manmade fertilizer use in the United States quickly doubled. Government policy subsidized and encouraged maximum grain output and the use of such chemicals. Tragically, these pro-production policies strongly discouraged permanent pastures, rotating crop fields into grass, diversified farming with animals, and the use of grass buffers. Faith in grass was replaced by a new belief in chemicals.

At the same time, agriculture was becoming more specialized and segmented. Laboratory-produced vitamin D and antibiotics were making it feasible, for the first time, to restrict animals indoors round the clock. Outdoor habitats for farm animals no longer seemed necessary and were starting to be abandoned in favor of crowded confinement systems. Instead of buffering grass strips, tree lines, and hedgerows, farmers were plowing fencerow-to-fencerow. Chemicals were coming to be regarded as saviors for fertility, pest control, and weed suppression.

Turning away from grass-centered farming has robbed billions and billions of animals of decent lives. It has also had severe ecological consequences. Soils and waters have lost their stalwart guardian as millions of acres of grasslands have been plowed, and vegetated buffers and grass crop rotations have been all but abandoned. Today agriculture is the nation's leading source of water pollution and is a major air polluter, on top of its contributions to global warming.

Especially alarming is the prospect that agriculture's failure to safeguard our soils is jeopardizing our future capacity to feed ourselves. Food production depends on soils created by geological actions over millions of years. But American soils are being blown and washed away at rates far faster than the earth can create them. Scientists estimate that farming methods that leave bare soils exposed for much of the year have, over the past two centuries, caused the United States to lose 30 percent of its topsoil.[48] Other than overpopulation, soil erosion is the single greatest threat to humans' continued ability to produce sufficient food, according to Cornell University's David Pimentel.[49]

Soil erosion rates in the United States are actually "many times higher than official estimates," according to a 2011 report by the nonprofit

Environmental Working Group.[50] "Across wide swaths of Iowa and other Corn Belt states, the rich dark soil that made this region the nation's breadbasket is being swept away." Federal policies create perverse incentives that continue to drive farmers from conservation and toward plowing up their lands, the report states. These include the 2007 energy bill mandates for corn ethanol, and the abandonment of landmark soil conservation practices adopted in 1985.

Alarmingly, American plows are still shredding and stripping its native grasslands. Research published in February 2013 in the *Proceedings of the National Academy of Sciences* by South Dakota State University researchers found that between 2006 and 2011, farmers in the Dakotas, Minnesota, Nebraska, and Iowa had plowed up 1.3 million acres of native grassland to plant corn and soybeans. Using government satellite imagery, the researchers concluded that the region's land-use changes—up to 5.4 percent annually—parallel some of the worst deforestations taking place in the developing world. "Between 2006 and 2011, over a million acres of native prairie were plowed up in the so-called Western Corn Belt to make way for these two crops, the most rapid loss of grasslands since we started using tractors to bust sod on the Great Plains in the 1920s."[51] Much of this cropland expansion has been in response to federal incentives for crop-based "bio-fuels."[52] The Nature Conservancy has called grasslands the world's most imperiled ecosystem, noting (among other things) that their demise has serious consequences for climate change.[53]

Where, one wonders, is the public outrage? Audubon Society and Nature Conservancy, two serious environmental advocacy groups, have attempted for years to draw attention to this looming ecological crisis right in the heart of our own country. But to little avail. Images of trees being felled and burned, even in distant lands, have garnered far more resources and interest.

California grasslands are seriously at risk, too. Several years ago, I attended a gathering at Stanford University of biologists, rangeland ecologists, ranchers, and policy advocates. The meeting was to discuss and share information about ecological services provided by rangelands and the gravest threats to their continued existence. Presentations detailed ongoing conversions of rangelands to housing developments, commercial property, and vineyards. Lately, groves of almonds, an especially thirsty crop, are the primary concern. A typical example: A land deal by Trinitas Partners, a Silicon Valley–based private equity firm, proposed to plow 6,500 acres of "rugged eastern Stanislaus County land from grazing to almonds," according to the *San Jose Mercury News*.[54] Ranching, the group convening at Stanford recognized, is among the most important ways of protecting the grasslands that remain.

Even putting aside concerns over habitat destruction, water use, and soil erosion, current agricultural methods will no longer be viable in the not-too-distant future. Manufacturing agricultural chemicals (fertilizers, pesticides, and herbicides) is fossil-fuel-intensive. Fossil fuels are used both as a feedstock and to power the energy-intensive production processes. Confinement animal operations and feedlots heavily depend on inexpensive, abundant fossil fuels. They are mechanized and use row crop feeds, which require energy to produce and are often transported great distances. Fossil fuels are a finite, diminishing resource that many experts believe are already more than half expended. More urgently, as Iowa-based PhD soil scientist and dairy farmer Francis Thicke argues, the extreme oil price spikes predicted for the coming decade will render fossil-fuel-based agriculture impractical long before the world's oil runs dry.[55]

Thicke's assessment is shared by soil scientists Lal and Stewart. Their report states, "Intensive agriculture has always been dependent on the energy market because of the energy requirements to manufacture and deliver fertilizer to the farmstead and to operate labor-saving farm equipment." They note that the world experienced a foretaste of the problems to come in 2008 when crude oil prices tripled to about $148 per barrel in a six-month time span.[56]

Food production that can truly be sustained over time will take new policies and fresh ways of thinking. Tomorrow's farming cannot look like today's. Transformation is urgent. High-tech "solutions" are regularly touted as the answer to agriculture's woes. But rather than frantically grasping at the latest shiny technology or haphazardly reacting to future environmental and oil crises, the United States should be engaging *now* in an orderly and deliberate transition away from fossil fuels and toward grass.

Yes, grass.

To begin the shift, Congress should act forthwith to reinvigorate the 1985 soil conservation standards. It must also forge future farm bills that tie agricultural incentives to safeguarding natural resources. As the Environmental Working Group's "Losing Ground" report puts it, we should be "requiring farmers to protect soil and water in return for the billions in . . . subsidies that taxpayers put up each year." This means, among other things, grass rotations for croplands, use of cover crops, and reestablishment of grass buffers. It means fully funding the Conservation Reserve Program (CRP), which provides incentives for maintaining grasslands rather than converting them to cropland.

Agricultural subsidies and incentives should, across the board, support grass. Farmers should be rewarded for converting croplands to grasslands, rather than the other way around. They should be incentivized for using grass (and other) cover crops instead of chemicals, and for establishing and maintaining grass buffers. To reduce animal suffering, make safer and more nutritious food, protect water, and rebuild soils, all farm animals—but especially cattle and other grazing animals—should be moved out of buildings and feedlots and returned to grass. Federal and state farm dollars should assist in the transition.

Fertility and pest control from grass rather than chemicals comes from biological processes. As part of this revised approach, we must halt our war on insects. Former USDA entomologist and farmer Jonathan Lundgren is alarmed by the mass disappearance of insects. Insect diversity is essential to functioning ecosystems. "The insect apocalypse is very real," he warned in early 2020. "We have lost 70 percent of insect biomass in the last 20 years." This is largely the fault of current agricultural practices, he says. The use of agricultural chemicals is one major problem. Destruction of nature's complex systems is another. "Agriculture has become much, much too simplified." The solution, Lundgren advises, is creating farming systems that are more like nature—with variety and complexity. To rebuild this complexity, we need farm animals. Manure, for example, is habitat for many types of bugs. Research of cow pats in South Dakota has found 172 insect species, while Canadian researchers have identified 450. Dung beetles are particularly valuable. Lundgren calls them a *keystone species*. "They drive everything else that happens," he urges. "Nature abhors monoculture. Biodiversity is life."[57]

Agriculture founded on manmade fertilizers, insecticides, and fungicides destroys diversity and is based on externalized costs. It's like throwing your garbage over the fence into your neighbor's yard. It's looking the other way as we cause problems for humans and animals downstream, downwind, and down the food chain. People today want information and transparency. The old black box of food production won't cut it. Now is the time for food production with true cost accounting.[58]

Our food system should steward, preserve, and regenerate our natural resources. With grass, we can break away from the polluting, soil-eroding model. We can build fertility and keep away pests with minimal chemicals. With animals on grass, we can build healthier, heartier herds and flocks that provide deeply nourishing foods. We can reestablish beautiful landscapes and functioning ecosystems. All while we rebuild soils and sequester carbon. In short, we can raise animals and grow food in ways that are good for our bodies, our souls, and the planet.

Grass, not chemicals and mechanization, should be our food's foundation. This is a massive but necessary shift. Today's model of agriculture is extractive. It pollutes and wastes water as it deadens and erodes soils. Regenerative farming, in contrast, where grass is a key component, mirrors nature's unending, restorative cycles.

Farming based on grass might involve the United States raising fewer farm animals. It might entail additional labor (for more animal husbandry and for the planting of cover crops), which means more employment but higher production costs. It could also mean some short-term lost farm revenue due to refraining from cultivating every available acre. (It is unhelpful that large swaths of cropland are now rented rather than owned, which disconnects farmers' economic interests from the benefits of long-term conservation.)

Yet oil's finite nature makes moving beyond fossil-fuel-based farming necessary and unavoidable. And with a transition away from fossil fuels, cattle and other grazing animals will play a central role. Their bodies catalyze vital soil microbiomes and maintain essential grasslands as they efficiently, reliably convert the sun's energy to nourishing food and drink. Perhaps we will yet fully realize the stalled dream of an American farming founded on grass.

CHAPTER 3

Water

I T ' S   O F T E N   S A I D   F U T U R E   W A R S   W I L L   B E   F O U G H T   O V E R   W A T E R , not oil. Whether or not this prediction proves correct, no one questions water's importance to human survival. The human body is 60 percent water; we can last just three days without it. Food production is likewise entirely dependent on hydration, often by irrigation. A truly regenerative food system wisely stewards water, keeping it pristine and plentiful.

Yet it is widely recognized that much of modern agriculture is a poor caretaker of our waters. Every year, new books, articles, and studies document industrial farming's contamination and waste of water. The federal Environmental Protection Agency has declared that agriculture is the United States' single largest water polluter. EPA blames US agriculture's sediments, nutrients, pathogens, and chemicals for 60 percent of impaired river miles and half of all polluted lake acreage. Runoff from farm fields, primarily crop production in the Mississippi River watershed, is responsible for a dead zone the size of New Jersey in the Gulf of Mexico.[1]

Beef production is sometimes singled out as a water polluter or, more frequently, as a heavy user of water. Have you heard that producing a single pound of beef takes a gargantuan volume of water? I have, countless times. Frequently the statement is followed by the conclusion that in today's ever-thirstier world, there is no longer room for cattle. I am sure this was among the reasons I stopped eating beef myself. Yet we've already seen how cattle's impacts on soils and climate vary dramatically, depending on how they are raised. When it comes to water, once again, "it's the *how*."

When I worked as an environmental lawyer specializing in water quality, such issues came into sharp focus. After graduating from University of Michigan Law School, I began my legal career as an assistant district attorney in Durham, North Carolina, then returned to my hometown, Kalamazoo, Michigan, working in private law practice.

After five years, I took a job at the National Wildlife Federation, in Ann Arbor, Michigan, an office focused on protecting the Great Lakes. Most of my time was spent advocating for strengthening and enforcing water-quality laws and regulations.

In 2000, Robert F. Kennedy Jr. offered me a job in New York at Waterkeeper. It's a collection of local groups, each working to protect a particular body of water. When I joined, Bobby had been hearing from many of these "keepers" that agriculture, especially the industrialized hog and chicken sector, was the biggest pollution problem in their regions. Algal blooms, fish kills, beach closings, and tainted groundwater caused by spills and leakages from confinement animal facilities were all becoming common occurrences. The Clean Water Act explicitly covers "concentrated animal feeding operations," yet the law was barely enforced, at either the state or federal level. My assignment was to change that.

Bobby charged me with launching a "campaign" (as he put it) to fight the pollution from concentrated animal agriculture. Not convinced I would enjoy spending all my waking hours focused on manure, I hesitated. But then I traveled to communities in Missouri and North Carolina overrun by industrial hog operations. I met people who'd spent their entire lives in the same home yet could no longer sit on their front porches or hang laundry outside due to the pervasive stenches. These were hardworking rural folks with little money or political clout.

I saw lifeless automated facilities blighting otherwise bucolic landscapes with their metal walls and concrete floors. They looked more like prisons than farms. Row upon row of these unwelcoming structures held thousands of sentient creatures. And I saw, and *smelled*, giant festering ponds of liquefied manure brewing behind every cluster of buildings. Manure was leaking and spilling into local streams and rivers, now choked with algae, and periodically filled with dead and diseased fish.

Community members—many of them farmers—had availed themselves of every means of objecting: writing letters, speaking at public meetings, protesting, even pleading with elected officials. But their elected representatives were turning a deaf ear and providing no help. I quickly realized I wanted to be part of this fight.

What I was witnessing was the results of a radical reshaping of farming that started mid-20th century. Small flocks and herds of chickens, turkeys, and pigs had long lived with humans. The animals spent their days roaming and foraging outdoors, with bedded barns offering nighttime protection and comfort. They were part of nearly every farm, broadly dispersed throughout the country. But spurred by the same shift in government policies that caused the abandonment of grass, together with new drugs and technologies, a handful of motivated entrepreneurs brought farm animals off the land and into crowded metal buildings. Particularly where land was inexpensive, concentrated animal operations sprang up. Grazing and foraging were replaced by concentrated feeds grown and transported, often from distant venues. North Carolina's confinement swine industry, which became the nation's second largest (after Iowa), was built on daily trainloads of corn shipped from the Midwest. The businessmen pushing hog and poultry farming toward the confinement model were all heavily invested in the feed sector.[2] Dispersed ownership of livestock and poultry by millions of independent farmers and ranchers dissolved as a few dozen agribusiness corporations gained ever-greater control.

Crowding, stress, boredom, and lack of sunshine, fresh air, and exercise soon resulted in skyrocketing rates of illness and death in poultry flocks and swine herds.[3] It therefore became routine to add antibiotics to daily rations, drugs that kept the miserable creatures alive in otherwise unlivable conditions. As an added bonus, USDA research showed that antibiotics sped up animal growth. The agency actively promoted the drugs to farmers, urging them to get animals to market faster. Adoption of these practices was enabled by lack of public awareness, misguided public policy, and lax government oversight.

My job at Waterkeeper was to engage in full-time combat against misuse of antibiotics and the pollution of the industrial animal sector. I was to provide assistance to activists, bring public attention to the crisis, sue polluters, and press for improving and enforcing statutes and regulations. I welcomed the challenge. Traveling the country, investigating the downstream effects of manure spills and leakages, I met and interviewed many activists and experts. Ordinary citizens were working alongside water biologists, epidemiologists, and sociologists, all meticulously documenting debilitating impacts on people, animals, and the environment. Along the way, I read thousands of articles and studies about water, soil, and air contamination and human physical and mental health issues connected with animal confinement operations. We began gathering and organizing people living near the facilities, providing them information and connections with fellow activists and experts. Soon my colleagues and I started preparing lawsuits against large hog operations in North Carolina.

In my legal research, I discovered that the profound changes to livestock and poultry farming were actually reflected in the language of the landmark federal Clean Water Act of 1972. Congress understood that the new style of industrialized facilities (concentrated animal feeding operations, or CAFOs) posed a novel threat to water in ways that traditional grass-based farming never had. Kansas senator Robert Dole made the following statement in the *Congressional Record*:

> *Animal and poultry waste has, until recent years, not been considered a major pollutant. The picture has changed dramatically, however, as the development of intensive livestock and poultry production on feedlots and in modern buildings has created massive concentrations of manure in small areas. The recycling capacity of soil and plant cover has been surpassed . . . The present situation and the outlook for the future developments in livestock and poultry production show that waste management systems are required to prevent waste generated in concentrated production areas from causing serious harm to surface and ground waters.*[4]

Soon the campaign I was leading for Waterkeeper started suing polluting confinement swine operations under the Clean Water Act, along with other environmental and nuisance laws.

My favorite part of the job was visiting well-run traditional farms and ranches. It filled me with pleasure and relief to see animals leading normal, natural lives. Animals frolicked in grassy meadows, jockeyed for position at water troughs, lay in the sun. These were the farm images I'd held in my mind since childhood. The farmers and ranchers caring for these animals spent much of their time outdoors. They deeply understood local climates and ecology. Cycles of regeneration were plain to see: Animals grazed fields; their urine and manure went directly onto the ground; feed crops and bedding would later be grown in those same fields. Wildlife and plant life were bursting forth at every turn.

But those visits to regenerative farms were rare treats. Mostly, I was fully immersed in tackling the industrial agricultural complex. It felt rewarding and invigorating to be doing such important work. Yet it was also disheartening. Obstacles to reform seemed nearly insurmountable. Political powerhouses—the food, chemical, agribusiness, and pharmaceutical industries—were all heavily invested in maintaining the status quo and aggressively fought off any efforts to get them to change their ways. Industrial animal operations were ugly, stinking, uninviting places. Locked away inside the buildings, workers and animals breathed fouled air, and creatures were packed together, manure-stained, and listless. Putrefying mountains of poultry litter or festering lagoons of liquefied porcine manure were surrounded by 12-foot-high chain-link fences with red-lettered signs blaring KEEP OUT!

These facilities are not merely off-putting, they're a constant threat to clean water. Any gardener knows the value of manure for growing plants. And on a traditional farm with a moderately sized flock or herd of healthy animals, manure is an irreplaceable, life-sustaining asset—economically and ecologically. As we've seen, returning manure to the earth keeps soils alive and generative; it reduces erosion, expands the ground's capacity to hold water and carbon. Animal manures enable cultivation of crops, including animal feeds, year after year.

"Livestock manure, when properly used, is a valuable resource that increases the biomass and biodiversity in agricultural systems," notes the Pimentels' textbook *Food, Energy, and Society*.[5] More specifically, manure application enhances biodiversity of soil-dwelling organisms. Studies in the former USSR showed that species diversity of larger soil critters (like worms and insects) increased by 16 percent when organic manure was added to experimental wheat plots. Organic manure added to grassland plots in Japan caused species diversity of soil organisms to more than double in number. Adding manure caused a 10-fold increase in soil organism diversity in Hungarian agricultural land.[6] Diverse, vibrant soil communities mean fertility, water, and carbon sequestration.

"Livestock manure is of tremendous value in holding the soil," notes Wes Jackson, founder of the Kansas-based Land Institute. "Its spongy nature absorbs the blows of rain and the water itself." He cites research in Iowa showing that 16 tons of manure applied to a 9 percent slope reduced the soil loss by over 17 tons.[7]

Yet as Senator Dole suggested decades ago, industrial livestock and poultry operations transform manure from ecological benefactor to pollutant, and from economic asset to liability. Much of the problem is concentration, which, in industrial operations, is extreme. While a typical flock was once a dozen hens, an industrialized facility will often have over a million. Where a hog farm traditionally had two dozen pigs, many confinement operations now have tens of thousands. With such large animal populations in such small spaces, keeping the facilities in balance with the environment is impossible.

The waste from industrial operations is tainted and infected. Jamming so many animals into buildings ensures rampant disease problems, correspondingly high levels of continual antibiotic dosing, drug-resistant pathogens, and massive waste output. The work of North Carolina State University water scientist Dr. JoAnn Burkholder has documented that typical confinement hog effluent contains hundreds of distinct contaminants, including heavy metals, hormones, pesticides, and pathogens.[8]

This voluminous waste is inherently problematic. Wastes are stored in huge stinking mounds (for turkey and chicken operations) and in festering football-field-sized lagoons of liquefied manure (for hogs, dairy, veal, and egg operations). The giant manure stores frequently leak into groundwater and run off into surface water, and constantly give off pollutants to the air. Leaching from these lagoons is so ubiquitous that state laws typically even expressly allow a certain amount. In Iowa, for example, a 7-acre lagoon is allowed to leak up to 16 million liters of untreated waste each year. Despite this allowance, a study of the state's lagoons found more than half of them leak in excess of the legal limit.[9] When manure storage structures are emptied, the animal waste is dumped on nearby lands, often overwhelming the absorption capacity of plants and soils, leading to further water pollution from contaminated runoff.

The other side of the water pollution problem comes from inputs to confinement operations. Of course, continually confining animals means they cannot graze or forage. This is an animal welfare concern, and it produces less nourishing food for humans. It is also an environmental problem. All of the feed must be delivered. In the United States today, some 55 percent of grains are destined for such operations.[10] Corn and soy farming entails plowing, planting, irrigating (sometimes), harvesting, drying, and transporting. Crop production is energy-intensive, the main cause of soil erosion, and the primary source of agriculture's greenhouse gases.

In the United States, "[c]orn production causes more soil erosion than any other crop," according to Professor David Pimentel.[11] Irrigated crops divert streams and suck up groundwater reserves. "In addition," the text states, "corn production uses more herbicides and insecticides than any other crop in the United States, thereby causing more water pollution than any other crop produced and therefore is a major contributor to ground water and river water pollution."[12] Corn farming is also by far the largest user of chemical fertilizers. Some 97 percent of corn acreage is grown with manmade fertilizers,[13] according to recent government data. The Gulf of Mexico Dead Zone starkly illustrates the enormous water pollution problem caused by American crop production, especially corn and soy.

Pigs, turkeys, and chickens are omnivores. They must be given some sorts of feeds. But cattle, sheep, goats, and bison can be raised simply grazing, with little or no added feeds. Mounting research shows animals raised wholly grass-fed produce healthier foods and provide more eco-system services.

In 2011, USDA scientists investigated the benefits of grass-based dairies.[14] The researchers created detailed models for four types of dairies—ranging from total confinement to fully pasture-based. For each dairy type, they estimated all major environmental variables, including air pollution, soil erosion, and water contamination. They also estimated emissions of the major greenhouse gases—carbon dioxide, methane, and nitrous oxide—from both primary production and secondary production of pesticides.

The models showed grass-based dairies as better for the environment in every regard, especially for soils and water. When high-producing dairy cows were kept in barns year-round, the associated sediment erosion from growing corn and alfalfa for feed averaged 2,500 pounds per acre. But with cows foraging on perennial grasslands, their diets supplemented as needed with purchased feeds, sediment erosion dropped 87 percent. Runoff of phosphorus, a major water pollutant, dropped 25 percent.

Not surprisingly, the grass-based farms also sequestered much more carbon. "When fields formerly used for feed crops were converted to perennial grasslands for grazing, carbon sequestration levels climbed from zero to as high as 3,400 pounds per acre every year." Cropland transitioned to pasture "can build up lots of carbon in the soil and substantially reduce your carbon footprint for 20 to 30 years," the study concluded.

True cost accounting for food looks at all inputs, outputs, unintended consequences, and downstream effects. This holistic approach is the only sensible way to evaluate a food's ecological footprint. Under true cost accounting, sequestered carbon offsets generated carbon. Recent research on well-managed beef operations, like White Oak Pastures, shows the potential to be better than "carbon neutral."

But even without considering carbon sequestration, total emissions for the greenhouse gases methane, nitrous oxide, and carbon dioxide were 8 percent lower in the grass dairy system than in the confinement system. Ammonia emissions were 30 percent lower. The researchers also found that keeping cows outdoors helped reduce fuel use and the resulting carbon dioxide emissions from farm equipment, because producers didn't need to plant and harvest as much feed for their livestock. "Average net farm greenhouse gas emissions dropped about 10 percent by keeping the herd outdoors year-round."

Calculating the overall carbon footprint for every pound of milk produced in each of the four systems, they found that a well-managed dairy herd kept outdoors year-round left a carbon footprint 6 percent smaller than a typical confinement herd. This was despite the fact that the confinement operation's cows produced milk at much higher rates than the grass operations, which had smaller cows. "Although the confined cow produced 22,000 pounds of milk every year and the foraging cow produced only 13,000 pounds, the total amount of milk protein and fat produced on the two farms was essentially the same, because the foraging cows produced milk with more fat and protein. In addition, the same amount of land supported a larger number of the small-framed Holstein/Jersey crossbred cows."

The shift toward animal confinement systems has gone hand in hand with the replacement of humans by machines and various technologies. From 1945 to 1994, as farm and ranch labor was displaced by machines and chemicals, farming's total energy use increased by more than 400 percent.[15] Fossil fuels go into agriculture directly and indirectly. About 60 percent of agriculture's energy goes to powering mechanized equipment and vehicles.[16] In feed production, this means vehicles and mechanized devices for plowing, planting, applying chemicals, harvesting, and drying crops. In animal agriculture, it also includes automated waterers and feeders, waste flushing and ventilation systems, and manure spreaders for dealing with the prodigious waste.

The other 40 percent of energy in US agriculture is used indirectly, as both the feedstock and the fuel to manufacture agricultural chemicals.[17] Manmade fertilizers were invented in 1830 but barely used for a century. This changed at the end of World War II when munitions factories were converted to chemical fertilizer plants. Between 1939 and 1945, fertilizer use in the United States doubled. Globally, agricultural chemical use skyrocketed. "Since about 1950 when the availability of fossil energy became readily available, especially in developed nations, this supported the 20- to 50-fold increase [worldwide] in the use of fertilizers, pesticides, and irrigation."[18]

In 2004, US farmers put 23 million tons of chemical fertilizers and 1.1 billion pounds of herbicides, insecticides, and fungicides on US soils. The 2014 Census of Agriculture shows that US agricultural operations used commercial fertilizers on 247 million acres; chemical herbicides on 285 million acres; chemical insecticides on 100 million acres; chemicals against crop disease on 35 million acres; and chemicals against nematodes on 14 million acres. As noted earlier, among US crops, corn farming is by far the largest user of agricultural chemicals.[19]

Agricultural chemicals have numerous downsides, such as reducing glomalin levels in soils. They also cause a lot of serious water contamination. The most comprehensive analysis of water quality in the nation's streams and surface waters was a decade-long study conducted by the US Geological Survey from 1992 to 2001. The startling results show "at least one pesticide was detected in water from *all* streams studied."[20] Although banned in 1992, DDT (along with other persistent pesticides) was also found in fish and bed-sediment samples from most streams and in the flesh of more than half of the fish. "Most of the organochlorine pesticides have not been used in the United States since before [this study] began, but their continued presence demonstrates their persistence in the environment," USGS reported.[21] Looking at groundwater, USGS found pesticides in both agricultural and urban areas. In shallower wells, "more than 50 percent of wells contained one or more pesticide compounds."[22] Contaminated wells are particularly worrisome for human health because rural people often have no alternative source of water.

Less understood are the ripple effects of agricultural chemicals on soils. What is clear is they disrupt the complex web of biological transactions. Soil scientist Dr. Kristine Nichols, while working at USDA, demonstrated that commercial phosphate fertilizer hampers glomalin production. Dr. Nichols has also warned that chemical insecticides and fungicides indiscriminately kill beneficial soil organisms.[23] We know soil's microorganisms are essential to its functioning, including water holding, fertility, and carbon sequestration. There can be little doubt that for soils to function optimally, the use of agricultural chemicals should be minimized.

Industrial animal production itself has also directly caused huge amounts of water contamination, including by pharmaceuticals. In 2011, the US Food and Drug Administration reported that 80 percent of all antibiotics used in the United States every year go to America's livestock and poultry. Of that 80 percent, FDA said, over 90 percent is given to animals that are not sick. Instead, the drugs are added to feed and water to lower feed costs (since antibiotics hasten animal growth) and to stave off diseases in crowded conditions. Yet up to 75 percent of antibiotics fed to animals pass unchanged through the animals' bodies and into animal feces and urine, entering the environment in full force and effect.[24] This raises the specter of broad effects on wildlife and humans who come in contact with the drugs in the environment and in meat. Medical and public health organizations have long warned that such antibiotic use is contributing to the worldwide rise of antibiotic resistance.[25]

USGS's landmark surface waters study found widespread contamination of water by antibiotics. The report noted that "unlike pesticides, which are intentionally released in measured applications, . . . pharmaceutical residues pass unmeasured through wastewater treatment facilities that have not been designed to deal with them."[26]

For pigs, chickens, and turkeys, some types of water pollution can be lessened. Well-run operations, with healthy living environments, for instance, need not feed antibiotics. But because these animals are omnivores that cannot survive on grass alone, water pollution from feeds cannot be eliminated entirely. Grazing, fresh air, and exercise benefit pigs and poultry immeasurably. But even on the best pasture, they still need to be fed.

Regenerative farms raise omnivorous animals as integrated parts of diverse operations. Pigs and poultry spend their days foraging and grazing, supplemented with feeds, crop residues, and surplus human foods. Such an approach reduces pollution from feed production, prevents antibiotic overuse, and provides animals happier, healthier lives.

Cattle, meanwhile, offer an even more enticing opportunity. Feeds—with all their attendant problems of fossil fuel use, soil erosion, greenhouse gases, and chemical pollution—can be avoided altogether. We know cattle have the special capacity to live on a simple diet of grasses and forbs. We've seen that nothing provides better protection to water and soils than a dense grass covering. We've also seen how cattle grazing makes healthier soils, grasses, and grass ecosystems.

Cattle, of course, can nourish themselves from their own foraging. This is why they (and other grazing animals) have always traveled with nomadic peoples around the globe. A bovine can be raised without any resources, economic or environmental, spent on plowing, planting, harvesting, drying, or transporting, nor any of the pollution that results from each of those steps. This unique attribute makes cattle vital to our global food supply, especially one designed for long-term sustainability.

Feeding farm animals can be environmentally costly. But grazing cattle give back even more than they get. The travesty of industrialized agriculture is that it robs cattle of this, their greatest contribution to the human food system.

The confinement mega-dairy, then, must be regarded as among the most offensive farming systems. Like the industrial swine operation, it deprives animals of sunshine, freedom of movement, and grazing. And it channels enormous volumes of manure into festering storage ponds that leach to groundwater, can spill to rivers and streams, and emit ungodly odors and pollutants to the air. The soy- and grain-based feeding regimens raise the same pollution and resource concerns as pig and poultry facilities. Worst of all, this bundle of problems is being foisted onto the backs of grazing animals—creatures that indisputably belong on grass.

The American dairy sector, like pork and poultry, has radically changed over the past century. It has shifted from small-scale and grass-based to large feed-based. In the 1930s, dairy cows were still present on 70 percent of farms, and herds were small—tiny, compared with today's. In 1945, milking herds in the leading dairy state, Wisconsin, averaged just 15 cows. Even by 2002, the state's average herd was only 71 cows. But dairy cattle everywhere were being bred for ever-greater volumes of milk production. And the bulk of the animals were moved from lush pastures of the Upper Midwest to drier western states where cows are continually confined, or may only have access to a small dirt or concrete lot. By 1994, California became the nation's leading dairy state with most milk coming from large confinement operations. And by 2006, California had more than 1,000 dairies having more than 500 cows. Today many operations have more than 10,000 cows.[27]

As with the rest of agriculture, the dairy sector's rupture with grass is directly tied to mid-20th-century historical events. "[I]n the decades following World War II farmers found that the use of relatively cheap energy, fertilizers, and pesticides, and greatly improved mechanization could improve farm profits," notes an article in the *Journal of Dairy Science*.[28] "These inputs, which allowed greatly increased production per cow, were substituted for pasture in the production process," states a Pennsylvania State University report on dairy farming history. "This trend gave rise to the predominance of confinement dairy practices on farms throughout the United States."[29]

Sadly, confinement dairies cannot provide cows good lives. When you consider a ruminant's innate daily patterns—walking, grazing, and masticating throughout the day—it's obvious continual confinement goes against their very nature. Concrete floors, lack of exercise, and restrictive physical space have led to an epidemic of lameness (35 to 56 percent)[30] among US dairy herds.

Lopping off a portion of the tails of dairy cows, "tail docking," is also common in industrial herds. It does make milking a bit easier. But it's not without costs. The cow loses her ability to swish away flies. And "there is evidence of acute pain associated with docking," a study in the *Journal of Dairy Science* states. "Moreover, tail docking may result in chronic pain. Sectioning the nerves in the tails of cattle can result in neuroma formation; neuromas can cause chronic pain, similar to the phantom pain felt after limb amputation."[31]

Due to the volume of liquefied manure they produce, confinement dairies are also inherently risky to the environment. To cite just one example, in 1998, in Washington State, manure lagoons at two confinement dairies collapsed within a week of each other. Each dumped its entire contents—one spilling 1.3 million gallons of liquefied manure, the other spilling 700,000 gallons. Both spills ended up in the Yakima River.[32] Citizens in the Yakima Valley, an area known for its scenic landscapes and productive apple and cherry orchards, got mobilized. They organized themselves and learned to skillfully document odors and water pollution from confinement dairies. When I was working at Waterkeeper, they called on us for assistance.

Working full-time on problems like these from industrial animal operations led me to question the attitude I'd adopted years earlier toward cattle ranching and beef. The more I saw and learned about how dairy cows, pigs, chickens, turkeys, and egg-laying hens were raised, the more I regarded beef cattle as having the best lives of all animals in agriculture. My friend Paul Shapiro, who served many years as a senior vice president at the Humane Society of the United States, has said if your eggs come from an industrial hen operation, "It's better to eat a steak than an omelet." And increasingly, I was seeing why. The more familiar I became with modern agriculture, the more I began to see an irony in beef being the first meat I'd given up in college.

Unlike most other farm animals, beef cattle start their lives with their mothers, nourished by nursing and grazing. Most will continue to live in herds, on pasture or rangelands, for about the first year of their lives. Regardless of how they are raised after that, they will never be confined to cages or metal buildings, or kept continually on slatted floors or concrete. For those reasons alone, I now believe beef cattle have by far the best lives of any animals in the food system.

I'll say more about overgrazing later. For the moment, I will just remind the reader vast ruminant herds once blanketed the globe. In Allan Savory's words, most of the world's grazing lands are not being overgrazed, they are being *improperly* grazed and "over-rested."

Other than overgrazing, the serious concerns about beef production —including water use and pollution—generally relate to feedlots. Most beef cattle raised for meat in the United States (as opposed to breeding herds) are confined for the latter portion of their lives to dirt lots, commonly called feedlots. Cattle feedlots elicit similar concerns as other forms of industrial food production: too many animals, too concentrated, too much manure, all leading to animal welfare concerns, more disease spread, and odors.

Cattle in feedlots, like all cattle, pee and poop where they are standing. Pens are periodically scraped, manure is gathered and stored, usually in piles. The manure is transported to fields for land application. Some urine will end up in that manure, some in the air. Cattle feedlots, especially large ones, tend to reek to high heaven, and are often the source of huge insect pest problems for neighbors. These are serious and legitimate concerns.

Nonetheless, there are important differences from other types of animal facilities. Most important, beef cattle are never crammed into crowded, disease-prone, stinking buildings. While a steer's life in a feedlot is undoubtedly monotonous, and lacks grazing, at least he is outdoors, on soft ground, with some room to move about. From an environmental standpoint, it's an important difference that beef cattle feedlots neither liquefy manure nor store it in lagoons. Handled as a solid, manure poses much less risk of leakage to groundwaters and spilling to surface waters, and there's less volatilization to the air. That said, the larger the feedlot, the more animals and manure, and the greater the likelihood that any of these potential problems will occur.

Nowadays it's popularly claimed beef feedlots and feeding grain to cattle are new ideas. This is simply untrue. Neither early Middle Eastern or European nor early American cattle husbandry was strictly grass-based. According to *A Short History of Farming in Britain*, at least as early as the Middle Ages domesticated cattle were fed exclusively "hay and corn" (corn in the British sense, meaning "grain") during the winter.[33]

Moreover, in the newly forming United States, cattle typically grazed on grass from spring to fall then ate hay and grains in the weeks or months prior to slaughter. Describing the Ohio and Mississippi Valley regions of the mid-19th century, a history of US beef explains that cattle were kept until three or four years of age, at which point they were put into feedlots and fattened on corn. "Grass was still abundant and relatively cheap and constituted the sole feed during the summer and fall; while hay, either timothy or prairie, supplemented with a liberal allowance of shock corn, formed the winter ration."[34] By the 1870s, US beef cattle breeding herds were being kept in the open ranges of the Far West, whereas cattle fattening (largely on corn) was mostly taking place in midwestern feedlots. While it is frequently suggested these days that feeding corn to cattle is a post–World War II invention, nothing could be further from the truth.

For environmental reasons, and in my gut, I always prefer to see any creature, and especially grazing animals, roaming rolling, grassy hills. But I am unwilling to uniformly condemn cattle feedlots. I fully appreciate what my good friend Will Harris, a Georgia grass-fed cattle farmer, has colorfully stated about feedlots: "Sending cattle to a feedlot is like raising your daughter to be a princess and then sending her off to the whorehouse." Yet, having visited several well-run smaller-scale feedlots, it is clear to me they are not all equally problematic. When relatively small in size, where only mature cattle (not calves) are accepted, careful feeding regimens are strictly observed, and runoff is carefully controlled, a cattle feedlot can be acceptable for both animal welfare and the environment.

Comparatively little of my time at Waterkeeper was spent dealing with cattle-related problems. The reason is simple: Most beef cattle and many dairy cattle (especially in the Upper Midwest) still reside on grass. And when animals are reared on grass, they provide ecological services, and water pollution is minimal.

Despite the prevalence of feedlots in the United States today (the Census of Agriculture says there are 13,734, most quite small), it's still true beef cattle are generally raised similarly to how it's been done for millennia: foraging their own feeds from grasslands and being fed grain for fattening prior to slaughter.

Breeding animals—mother cows (which are 28.9 million of the 89.9 million cattle in the United States[35]) as well as bulls, most of which are in the western United States—typically spend their entire lives grazing rangelands or pastures. They are often moved, especially in colder climates, to higher elevations for some part of the spring, summer, and fall, then led down to areas near the home ranch, and supplemented with hay, over the winter. Many cattle will graze the stubble of field crops in the fall. Cattle mating and birthing (calving) happens with little human interference, usually out on the range. Mother cows have a single calf each year and have the satisfaction of providing their baby a mama's care. Beef calves have the pleasure and enrichment of being reared by their mothers, and stay with them, on pasture, for the first several months. Calves also typically stay on grass for some time after weaning, even if they eventually go to feedlots. The mother cows, bulls, calves, and yearlings are those black or reddish brown bovines grazing the hills and fields you see on nearly any American road trip.

The black-and-white-spotted cattle you see grazing are generally dairy heifers, meaning future dairy cows. In some parts of the country, especially in Wisconsin and Vermont, you will also see mature dairy cows out on grass. Most dairy heifers, though, once they are mature enough to enter the milking herd, are taken off grass and confined to buildings from that point on. There, like all animals raised under the confinement model, they are fed concentrated feeds like soy, oats, and corn, often mixed with some alfalfa hay.

The net effect of cattle in the food system is to benefit the world's waters. There is simply nothing better at absorbing, holding, and filtering water than lands densely covered by grass, and there is no better way to keep those grasslands biologically diverse, hydrated, and vibrant than by periodically grazing them. Grass-based farming systems also create filters for water going to rivers, streams, and groundwaters. Research in California's Sierra foothills has shown "grazing can actually enhance the ability of riparian vegetation to filter nitrate out of surface waters." In field experiments comparing test plots of grazed and ungrazed land in annual grassland-oak savanna, soil water from plots with grazing removed for only two years contained up to *five times more* nitrate than soil water from plots that were grazed. The reasons for this are not fully understood. Researchers hypothesize the effect may be due to reduced nitrate use by plants not subjected to grazing.[36]

I am not suggesting there have never been cases of water contamination linked to cattle on grass. My point is that well-managed grazing does not cause water pollution. The *overall effect* of having grazing cattle as part of the food system is a net positive for water—both for quantity and quality. Compared with nearly any other land use, and certainly with croplands, grasslands are far better at holding soils, making nutrients available for plants, keeping pathogens, sediments, and nutrients from entering groundwater and surface waters, and filtering rainwater. The Land Stewardship Project of Minnesota has even recommended permanent pasture for protecting waters, stating: "Permanent pasture for grazing livestock can be an ideal choice for minimizing water pollution."[37]

Fundamentally, water quality is better off with cattle than without them. A food system that includes cattle will be a better guardian of waters than one that excludes them. The more time the cattle spend on grass, the cleaner and more "effective" waters will be.

Quantity is the other side of the water question. In recent years, many people have come to accept the widely circulated idea that beef is particularly and unacceptably water-intensive. Vegan and environmental groups have made this a key point in their indictments of beef.

Anti-meat advocacy literature generally calculates water usage with an elementary arithmetic. The simple equation adds together three numbers —the volume of water a steer drinks, water used to grow feeds, and water used in slaughter and processing.

The most commonly cited figure (often used by media) is that it takes 2,500 gallons of water to produce a single pound of beef (in metric, about 20,820 liters of water per kilogram of beef). Occasionally, you'll encounter even much higher figures, such as 12,009 gallons per pound of beef. The latter number, said to originate from a book edited by Cornell's David Pimentel, is from a chart created for a vegan website. The vegan chart also says a pound of potatoes takes just 60 gallons of water and wheat takes 108 gallons. "As you can see from Professor Pimentel's figures," the website's commentary remarks, "it takes roughly 200 times more water to make a pound of beef than a pound of potatoes."[38] But is the math really this simple?

In a word, no. For starters, the foods selected for this chart seem deliberately chosen to create the impression beef is an extreme water guzzler. Potatoes and wheat are relatively water-thrifty. But calculations from other sources (unconnected to the meat industry) show that some of the most water-intensive foods eaten by humans include common vegan staples like rice, which takes 3,400 liters of water per kilogram (21 percent of the water used in global crop production), and sugar, an unnecessary (and unhealthy) food additive, which takes 1,500 liters of water per kilogram.[39] The vegan chart, not surprisingly, excludes these common pantry items.

Putting aside momentarily the vast nutritional differences among the foods on the chart I just mentioned, how believable are its water usage numbers? The chart looks legitimate, since it apparently comes from a textbook edited by Pimentel and other university professors. Although Pimentel's figures about energy and water in the food system are often touted in anti-beef writings, Pimentel himself was not an advocate for veganism. As a scientist, he openly acknowledged the limitations of such numbers. His writings state the beef figure was calculated from an animal that consumed 100 kilograms of hay and 4 kilograms of grain per pound of beef produced, and it assumes a generalization that 1,000 liters of water were used to produce 1 kilogram of hay and grain.[40] This is a big assumption. (Note: If you convert to metric, the wheat figure on the chart says it takes 900 liters of water per kilogram of wheat, but here, when calculating grain for the steer's consumption, the number was elevated to 1,000 liters.) Although this 1,000 liters appears to include rainwater, most people reading it and citing it will think it means irrigation water. If you exclude rainwater (which I believe you should), Pimentel's data would be wildly off for animals fed hay and grain that were grown *without* irrigation, which is the way the vast majority of US grain and hay is, in fact, grown.

Equally important, Pimentel implicitly acknowledged that the figure is inapplicable to cattle that are *not* fed grain and hay. Thus, the numbers have limited application. They tell very little about water usage for cattle raised entirely or mostly without feeds—in other words, cattle raised on non-irrigated pastures and rangelands, a common scenario in the United States and world today.

Clearly, we need to look to additional sources for the water usage figure. My scientific and legal training (and being reared by two educators) taught me long ago to always look critically at information and seek data from the most reliable sources. The beef industry has a great deal of data about cattle and beef, but is not always objective. For this reason, little of this book's information comes from the beef industry itself.

When it comes to analyzing water demands of beef, the most credible and thorough analyses have been done by agricultural colleges. While implicitly linked with the beef industry (and thus somewhat less objective), they are the most familiar with every life stage of the animal, the husbandry, and the processing.

In response to concerns over water conservation, researchers from University of California–Davis undertook the most detailed look at the water going into beef of any I've seen.[41] Whereas the water usage figures cited by advocacy groups are worst-case scenarios, the figure compiled by UC Davis is balanced. Rather than showing the least possible water usage, it strives to achieve a number that actually reflects *typical* water usage for a *typical* piece of beef. In the nine-page analysis, it goes, in excruciating detail, through every step of an animal's life, taking every variable into account. Among factors considered are where cattle were located, average ambient temperatures, the number of days on feed and feed types, typical grain types, how many acres of each type of grain would have been irrigated, and so on.

At the end of their thorough analysis, UC Davis researchers conclude it takes 3,682 liters of water to produce a kilogram of boneless beef (in other words, 441 gallons per pound). Interestingly, you may remember that's nearly the same amount of water it takes to produce a kilogram of rice. Having read UC Davis's full report, I find the number credible.

Taking this analysis a step further, if we specifically consider grass-raised cattle it becomes still clearer beef is not *inherently* more water-intensive than other foods. Author Lierre Keith, who returned to meat consumption after 20 years on a vegan diet, does a credible job breaking down this question in her book *The Vegetarian Myth*. She writes:

On pasture, beef cattle will drink eight to fifteen gallons of water a day. The average pasture-raised steer takes 21 months to reach market weight. That's 40,320–75,600 pounds of water total for an entire cow [sic]. That's 450–500 pounds of meat, with another 146 pounds of fat and bone trimmed off, which in an earlier, saner era would have been valued for food as well. Taking the mean of 475 pounds, the midpoint of 57,960 gallons yields a figure of 122 pounds [sic] of water per pound of meat [a figure that excludes the organ meats, the most nutritionally dense part of the animal].[42]

(Note that Keith's calculations show that it takes 122 gallons of water per pound of grass-fed beef. The word "pounds" is a typo in the book.)

Now, if we put nourishment into the accounting (and for such comparisons to be meaningful, we *must*), the idea that beef is overly water-intensive collapses completely. Keith adds nutritional value to the equation by looking specifically at the oft-made comparison with wheat. She points out that not only is beef much more nutritionally dense, but its nutrients are also far more usable by the human body (a point I explore more fully in the second half of this book). She writes:

The beef contains almost twice as many calories (592 v. 339, per 100 grams) . . . For wheat, sixty pounds of water produces 1524.45 calories, or 25.7 calories per pound of water. For grassfed beef, it's twenty-two calories per pound of water. And there's more than simple energy: those beef calories contain more nutrients, especially essential nutrients protein and fat. The numbers on those are 21g v. 13.7g and 8.55g v. 1.87g respectively. It's also crucial to understand that the protein in beef contains the full spectrum of necessary amino acids and is easy for humans to assimilate, while the protein in the wheat is both low-quality and largely inaccessible because it comes wrapped in indigestible cellulose.[43]

These figures show how misleading it is to treat all methods of cattle husbandry as the same. It's equally wrong to use worst-case scenario figures and pretend they are typical. Depending on where and how feeds are grown, and the type of feeds used, the water-usage numbers can vary dramatically. If you take out rainwater and use an average number instead of a worst-case scenario, the number plummets. Keith's analysis and the Pimentel figures both clearly show water usage in beef comes mostly from growing feeds. These numbers cannot be used for cattle raised on non-irrigated feeds, let alone cattle raised without feeds, on forages. The popular anti-meat internet memes about water-thirsty beef are inaccurate, and by a lot.

Nutritional differences between foods compounds the absurdities of these comparisons between different foods. As Keith's writings show, beef's nutritional content on a per-calorie or per-pound basis compares very favorably with wheat, as it does with other foods. Remember the work of sustainability professor Michael Lee of Rothamsted Research regarding climate impact of various foods? His analysis shows because beef is rich in essential, biologically available nutrients, when nutrition is considered, beef has a comparatively low greenhouse gas impact. The same is true for water impacts. If nutritional value is added into the comparison, it becomes plain a straight pound-to-pound comparison of beef with wheat or potatoes just doesn't make sense. Actually, it's quite a bit worse than comparing apples to oranges.

There is also another crucial aspect of water usage to consider: What happens to water *after* it hydrates farmed plants or animals? From a broader ecological perspective, the water a plant or animal uses in its growth is only one part of the impact it has on water resources. All water used in human endeavors ends up somewhere. In manufacturing, it is often discharged into a river or lake via tailpipe. In agriculture, water used to hydrate crops or animals will be released to the environment in diffuse ways. That water, too, through its presence or absence, and whether polluted or pristine, goes on to affect wildlife and groundwater, lakes and streams, and even air.

As shown earlier, much water landing on plowed fields is lost. It flows off, washing away precious topsoil and picking up contaminants along the way. Rainwater or irrigation water falling on fields of conventionally grown fruits, grains, and vegetables also becomes contaminated by agricultural chemicals, like pesticides, fungicides, and fertilizers. Water that runs off or evaporates is failing to hydrate soil and the living system it sustains. It's gone from the ecosystem. (This is Savory's "ineffective rain.")

On the other hand, water retained by plants and soils continues to provide life-giving hydration to surrounding organisms and thus supports the entire ecosystem. (Savory's "effective rain.") Water hydrating cattle on grass ends up watering grass as well as subsurface communities of insects and microorganisms. In other words, water cycling through the animal living on grass goes on to support many other life-forms in the grassland ecosystem. When comparing beef's water usage with other foods, such differences must be taken into consideration.

There is no better way to safeguard water in our food system than for rain to land on grass-covered earth. Research in the Midwest has shown that compared with cropland, perennial pastures used for grazing can decrease soil erosion by 80 percent and markedly improve water quality.[44] In fact, there is no better way to filter runoff than through healthy grasslands. University of Georgia research shows runoff rates of croplands are 5 to 10 times higher than those of perennial grasslands. Even more dramatically, erosion rates from croplands are 30 to 60 times higher.[45] Grass-covered lands benefit entire ecosystems by holding much more water than croplands.

Cattle enable grass to be a foundation of our food system. Whether grazing on rangelands, annual or permanent pastures, grass buffers, or cover crops, cattle make it possible to produce food from densely vege-tated lands that resist wind and water erosion, effectively retain water, and simultaneously support countless species, both above- and belowground.

In the global warming discussion I noted that, depending on farming methods, climate impact can vary by a factor of 10. In a similar fashion, the quantity of water used to raise a food is highly variable depending on how and where it's grown. Whatever plants or animals are farmed, water can be used wisely or wastefully.

Later, I'll describe in more detail how we raise cattle on our own Northern California ranch. Here's a quick overview of our approach, which uses water sparingly. Our cattle herd is grazed on sections of land in an area that totals about 1,000 acres. All of this land is natural grasslands. In another era, it was maintained by large herds of wild elk, pronghorn antelope, and deer. Prior to that, an array of large grazers and predators populated this land and maintained broad, open areas of grass.

Our climate is Mediterranean. Temperatures are moderate year-round, and nearly all rainfall comes during our coldest months—roughly from late October to early April. Summers are cool. It is windy year-round. (Keeping a baseball cap on can be challenging.) The terrain is rugged—hilly with deep gullies and ridges.

This weather and topography mean this land cannot (and should not) be used to grow crops. Yet it's ideal for grass. Grasses are not bothered by winds, fog, or cool temperatures. Many grasses here are known as cool-weather grasses, because they flourish in those cold, rainy months.

Our cattle are never fed grains and live year-round on grass. Naturally occurring vegetation is always their primary source of nourishment. We do no plowing, no planting, no irrigation, and no chemical applications to our land. We withdraw no water from wells or municipal water sources and divert no rivers or streams. The only water that hydrates the dense vegetation covering the ground of our ranch is rain falling from the sky.

We gather some rainwater in a retention pond. This is then distributed via a gravity system to troughs dispersed around the ranch. The water that goes to our home, and to the animals to drink, is collected rain.

Every time one of our cattle pees or poops, we celebrate. That animal's urine and manure fall directly onto our pastures, returning moisture and organic matter (as well as nitrogen, phosphorus, and beneficial bacteria!) directly to our grasses and soils. This, combined with the animal-impact pruning and smushing of biological materials into the ground, enables the grasses and other vegetation here to thrive. It is a cyclical, regenerative system that works well not only for the cattle but also for the hydrological cycle and the wild plants, animals, and fungi.

Every situation where cattle are raised is unique. Our ranch is a living display of land used to keep cattle without intensive water use or water pollution. We are proud of the way we care for our animals and our land. But we do not consider ourselves exceptional. I've been to dozens of other farms and ranches in many parts of the country where cattle are being raised appropriately and well. All of these farms and ranches conserve and protect water. In many regions of the United States and around the globe there is nothing better suited for the topography and climate, as a livelihood or a food source, than cattle tending.

All food production has impacts on water resources. Nothing about cattle raising is inherently more water-consumptive than other forms of agriculture. This is crystal clear when the nutritional value of the foods produced is taken into account. Looked at holistically, other forms of agriculture—particularly conventional farming using plowing and chemicals—are much more of a water pollution concern. Well-raised cattle, especially when raised on grass, help safeguard the world's water.

CHAPTER 4

Biodiversity

I'VE OCCASIONALLY HEARD IT SAID THAT BIODIVERSITY—THE degree of variation among an ecosystem's plants and animals—is harmed by the presence of cattle. This discussion will be quite brief because I don't think I need to vigorously refute this particular criticism. Even among people who seem to loathe beef, this notion has fallen out of favor. A growing body of research and real-world examples have shown well-managed grazing animals improve ecosystem function. The *Livestock's Long Shadow* report, amid a sea of criticism of cattle, states: "There is growing evidence that both cattle ranching and pastoralism can have *positive* impacts on biodiversity."[1] Indeed, there is now abundant scientific research showing that cattle grazing does more to help biodiversity than harm it. Although the public discussion of cattle and beef generally lags decades behind the research, on this issue it's more in sync.

Biodiversity includes everything from microscopic organisms to elephants. Earlier, I showed how grass and grazing enhance biodiversity of soil organisms. Anecdotally, cattle ranchers using soil-building methods, like those advocated by the Savory Institute, report seeing more biodiversity as they improve their soils. I've spoken with dozens of farmers and ranchers who witnessed more plant and animal diversity when they moved away from conventional agriculture toward regenerative. "We've got more wildlife than ever and more livestock than ever," cattle rancher Jerry Doan of North Dakota said, attributing this to cover crop mixing, more litter, and biodiversity starting at the soil level.[2]

Environmentalists often speak of "mitigating damage" from agriculture. But done well, cattle keeping is not about mitigation at all. It's an opportunity to create more diverse, more resilient ecosystems. Grazing can be hospitable to wild plants, fungi, and animals. In sharp contrast with plowing land and cultivating crops, the types of disturbances caused by cattle are like those the globe's ecosystems have experienced for millions of years. Paul Krausman, University of Montana wildlife conservation professor, has noted that some of today's greatest threats to wildlife are tillage agriculture, urban sprawl, tree and shrub invasion, and energy development. These "result in broad-scale loss and degradation of habitat that overwhelms management of remaining fragments."[3] Krausman also points to "a host of vegetation manipulations" that harm wildlife, particularly ground-nesting birds. In addition to agricultural tillage, these manipulations include "herbicide application, mechanical sagebrush removal, and over-prescription of fire in xeric landscapes."[4] These are land management practices often used as alternatives to grazing. "Wildlife managers in the Great Plains readily acknowledge the importance of livestock grazing to conservation," Krausman notes, "because ranchers whose operations remain profitable are less likely to convert native prairie to cropland."[5]

In contrast with those problematic land uses, "a number of studies have recorded higher biodiversity in grazed relative to ungrazed systems."[6] Just as Allan Savory has so well articulated, these studies show grazing is not only a beneficial but in fact a *necessary* "disturbance" in grassland ecosystems.[7] And as Savory also points out, because the world's wild herds have been largely extirpated, domesticated livestock must be their proxy. In the words of Nature Conservancy scientist Jaymee Marty, cattle can now "serve as a functional equivalent to large herbivores that historically grazed grasslands and play an essential role in maintaining biodiversity."[8]

Previously, I mentioned a University of Maryland field study with 40 test plots that found the least plant diversity on plots where grazing had been excluded. Notably, that study also documented the greatest beneficial effects on plant diversity with grazing by *large* animals, wild and domesticated: cattle, pronghorn, and elk on North America's Great Plains; wildebeests and impala on Africa's Serengeti Plain.[9]

Several earlier studies reached similar conclusions. A 2001 study in the journal *Plant Ecology* by Richard H. Hart, a senior range scientist at the USDA's High Plains Grasslands Research Station, describes field tests from several Colorado rangelands. The study sites had varying amounts of cattle grazing—no grazing, light, moderate, or heavy grazing—each practiced for over 50 years. Hart reports plant diversity was lowest where cattle had been excluded. He notes that this type of grassland, known as a shortgrass steppe, when moderately or heavily grazed by cattle "was similar to and probably as sustainable as steppe grazed for millennia by bison and other wild ungulates."[10]

A long-term study by researchers at University of Nevada, published in 2004, also found more plant diversity with grazing than in areas where grazing had been excluded.[11] In 1934, when the Taylor Grazing Act was adopted, multiple test sites that entirely excluded grazing were established. Sixteen of the sites remained when the Nevada study was conducted. To ascertain the effects of large grazing ungulates, the research team meticulously examined the ecologies both inside and outside the defined study areas. The researchers then subjected their findings to a thorough statistical analysis. They concluded that "light-to-moderate grazing in the Great Basin certainly has no ill effects on the ecosystem," and that the grazed areas had a greater variety of plants.[12]

As these field studies suggest, a number of real-world investigations into grazing's impacts have found the test plots from which cattle were entirely *excluded* the most revealing. A university textbook on grasslands confirms this: "A growing body of research shows that livestock grazing can enhance biodiversity. To a surprising degree, this research comes from cases in which, as part of conservation efforts, livestock grazing was removed, and subsequently, species or habitats of interest disappeared."[13] The unexpected results of such studies—that land or particular species' populations are further degraded without cattle—should no longer surprise anyone. (Of course, had the researchers asked Allan Savory, he could have told them all of this decades ago.)

A telling US example of such research outcomes involves California's vernal pools. These are ephemeral, shallow ponds that appear in flat, grassy areas during the rainy season (November through April). Attention was first brought to their ecological importance by a PhD botanist and San Diego high school science teacher, Edith Purer. After years spending her summers surveying them, in 1937 Purer presented a report on California's vernal pools to the Ecological Society of America. Only decades later was their uniqueness fully appreciated. In the 1970s, research of vernal pools cataloged more than 100 vascular plant species and more than 34 crustacean species. Astonishingly, half of the life-forms inhabiting them are found nowhere else on earth.[14]

Meanwhile, because some ranchers attempted to fill them, and cattle stepped into them, grazed their grasses, and drank from them, many environmentalists came to regard cattle ranching as a serious threat to vernal pools. "A major controversy surrounding the conservation of vernal pools concerns grazing," a Nature Conservancy document noted. "Some argue we need to remove cattle from the vicinity of these threatened habitats, and others argue cattle should be used to manage vernal-pool grasslands. In fact, cattle grazing was implicated as a major contributing factor to the decline of four vernal-pool crustaceans listed under the U.S. Endangered Species Act (USFWS 1994) with little to no supporting scientific data."[15]

After years of debate on ranching's impacts, the dearth of scientific evidence was finally addressed by The Nature Conservancy. Jaymee Marty, lead scientist for organization's Central Valley and Mountains region, began conducting detailed field tests to evaluate the effects of cattle ranching on the state's vernal pools. Her study involved 24 natural vernal pools, of varied sizes and depths, across two different soil types. On some study sites, grazing was unrestricted, on some it was seasonally restricted, and on some it was completely removed.

The findings of this careful, credible, non-industry-funded study shocked many in the environmental community. Marty reported "significant negative effects on the native plant community, pool hydrology, and the aquatic invertebrate community with the removal of grazing." Correspondingly, she found "that livestock grazing played an important role in maintaining species diversity." The report concluded: "Grazing should be considered one of a variety of important tools for land managers interested in the maintenance of biodiversity."

Jaymee Marty studied cattle grazing and vernal pool grasslands for ten years to determine the best approach to conservation and restoration of the pools. She was interviewed in 2009 about her unexpected (by some) findings. She summarized her major conclusions by saying, "I learned that grazing is important to maintaining these unique grasslands. Grazing keeps the non-native grasses from crowding out the native plants and sucking up all the water."[16]

Marty's study write-up for The Nature Conservancy emphasizes the importance of grazing animals in how ecosystems evolved. Wildlife conservation professor Krausman echoes the idea that large grazers have been essential to the functioning of the globe's ecosystems for millions of years. "Indeed, many, if not most, ecosystems rely on grazing by native ungulates to influence vegetation structure and composition," he writes.[17] Because California's ecosystems evolved with large grazing animals, the ecosystems need these animals. The role of cattle herds is equivalent to historic grazing herds, helping ecosystem function. Marty explains:

> *Why is cattle grazing so clearly beneficial to biodiversity in these vernal pools, when conservationists often advocate severe grazing restrictions? The answer to this question comes in part from the fact that California grasslands have a long history of extensive grazing dating back to the Pleistocene but were most recently grazed by herds of tule elk (*Cervus elaphus nannodes*) and pronghorns (*Antilocarpa americana*) before livestock introduction in the late 1800s. Hence, the pool species are adapted to some level of grazing. In addition, the plant species composition of California Central Valley grasslands has changed significantly since European settlement and is now dominated by exotic annual grasses. Thus, a long history of grazing coupled with the altered plant community yields a system that is now adapted to the changes brought about by cattle and one that becomes quickly degraded when cattle are removed.*[18]

Similar findings have been made by scientists studying land management approaches in the federal Conservation Reserve Program. The program allows some, highly restricted, grazing on areas taken out of crop production. In addition to the soil improvements previously discussed, Natural Resources Conservation Service field scientists have found grazing benefits a host of animal and plant species.

Kevin Reynolds, district conservationist for the NRCS in Decatur County, Iowa, has seen, year after year, the positive effects of grazing compared with fallowing ground without grazing. He notes, especially, the value for songbird populations. Grazing, Reynolds says, reduces thatch that makes it difficult for young birds to get around in older stands that haven't been disturbed. Properly timed and controlled grazing, Reynolds has observed, leads to better overall wildlife habitat. It also protects water and soils, and controls woody vegetation. In other words, grazing acts as a catalyst for the entire ecosystem. "The year after grazing, I've seen a noticeable difference in the number of bugs and in the number of birds that eat them on [this] land," Reynolds says, adding, "I'd like to see more grazing here."[19]

Other studies of CRP land have shown taking land out of crop production and allowing grasses to return helps many wildlife species. Grassland birds are declining more than any other bird group in North America. Converting from crops to grass has helped protect and reestablish a number of wild bird species, including the bobwhite quail, Bachman's sparrow, short-eared owl, ovenbird, acorn woodpecker, and greater sage grouse.[20] In a similar vein, Austrian studies have shown grazing improves habitat for wading birds.[21]

Several bird experts in our region have told me they once worried about cattle but, over time, they have become strongly "pro-cow." Their own fieldwork has taught them that grazing, by maintaining healthy grasslands and keeping open spaces open, provides invaluable assistance to many types of birds. Some of the birds that benefit from the reduction of thatch on our ranch include the hunters—herons, egrets, hawks, and falcons—to name just a few. Many bird experts and rangeland scientists now believe, "What's good for the herd is good for the bird."

A 2019 article by two such scientists, a rangeland entomologist at Montana State University and a wildlife biology professor at University of Montana, is a good example. It begins by noting, "Land use such as livestock grazing—the most common use of rangelands—influence the abundance and composition of insects, which may have far-reaching effects on rangeland ecosystems. Grazing impacts arthropods through direct habitat disturbance as well as by changing the composition and physical structure of plant communities they rely upon." The article discusses Montana research that compared "grazed, rested, and idled" grasslands. "The types of insects that provide a critical food source for sage grouse chicks and other shrub- and grassland-dependent birds were 13% more prevalent on managed versus idled rangelands," they write. And, they note, research shows that grazing strategies can be an effective tool for maintaining arthropod biodiversity. Researchers from Montana State University investigated abundance and diversity of ground-dwelling arthropods in sagebrush habitats in Montana over a four-year period. Interestingly, total insect catches were twice as high on idle pastures compared with managed pastures. However (and this is a big however), the specific insect classes preferred by sage grouse were more prevalent on managed pastures, and, even more important, managed rangeland resulted in a "more diverse assemblage of ground-dwelling insects." This, the scientists noted, is likely to be "particularly beneficial for birds."[22]

Grazing also helps other types of wildlife, including rare species, a phenomenon documented by a wealth of field research. Grazing boosts these species by altering grassland structure to create shorter vegetation, more openings, and more structural heterogeneity in general than is found when livestock are excluded. Specifically, studies have shown grazing helpful in reviving populations of swift fox, Karner blue butterfly, gopher tortoise, Louisiana black bear, eastern collared lizard, and salmon.[23] Other studies have found that livestock grazing supports a variety of beetles, kit foxes, kangaroo rats, blunt-nosed leopard lizards and other lizards, ground squirrels, and San Joaquin antelope squirrels.[24] The very existence of grazing areas supplies essential open habitat to countless wild plants, fungi, and animals.

Pollinators, indispensable to both wild ecosystems and the human food system, are an important example. According to the US Fish and Wildlife Service, over 75 percent of flowering plants—including apples, almonds, and squashes—need an animal pollinator.[25] Globally, about three-quarters of crop species depend on animal pollinators to produce fruits or seeds, and over 90 percent of vitamins A and C in foods are derived from pollinator-dependent crops.[26] Yet domesticated bee populations have been plummeting for years, suffering from a phenomenon called colony collapse disorder (CCD). Statistics released in March 2013 showed a 40 to 50 percent decline in bee colonies over the preceding year.[27] Numerous studies now strongly support the conclusion that domesticated bees are being killed off by a certain class of pesticides (neonicotinoids). The European Union has banned neonicotinoids for that reason. The causes of colony collapse are still debated.

What's certain is, especially in light of struggling domesticated bee colonies, wild pollinators have enormous significance for our food supply. About one-third of crop pollination is done by non-domesticated pollinators.[28] And nothing is more important to the survival of those wild pollinators than preservation of their rapidly disappearing habitat.

Just where do these wild pollinating animals make their homes? For many: rangelands. The very same rangelands that exist because they are grazed by cattle. "Preserving rangelands has significant economic value, not only to the ranchers who graze their cattle there, but also to farmers who need the pollinators," says biology professor Dr. Claire Kremen. She and her fellow UC Berkeley researchers conducted a study of wild pollinators' significance to the food system.[29] Their research shows humans need wild pollinators and wild pollinators need cattle. Professor Krausman's research reinforces the important role of grazing animals in protecting this habitat. He writes, "the use of rangelands for sustainable livestock production has the potential to ensure the maintenance of wildlife habitat, especially when compared to energy development and urbanization . . ."[30]

Researchers from Stanford University have urged the design of future food systems with wildlife in mind. Rather than focusing on wildlife preserves, they argue, farms and ranches should be part of the strategy for safeguarding the world's wildlife. "Countryside biogeography is an alternative framework, which recognizes that the fate of the world's wildlife will be decided largely by the hospitality of agricultural or countryside ecosystems."[31]

All human land uses affect plants and animals. The types of effects will depend on how those human activities are carried out. Of course, none of this research shows that every ranch or farm raising cattle fosters biodiversity. For example, ranchers filling in vernal pools are harming those ecosystems. But that's an example of poor land stewardship. It's not an indication ranching per se poses an environmental problem.

The reverse is actually much closer to the truth. Research now confirms having cattle on the land creates and sustains habitats and whole ecosystems, benefiting countless plants, fungi, and animals. And it shows that well-managed cattle ranching can have a significant positive effect on the world's natural diversity. Research like The Nature Conservancy's vernal pool studies are deepening understanding of how humans can work *with* nature instead of against it. The disappearance of the world's wild grazing herds created a gaping ecological hole. The best ways of managing cattle treat them as filling that void. Herds of cattle—large grazing animals over which humans have substantial control—provide invaluable nutrition to humans and present us the opportunity to protect and even re-create highly functioning ecosystems.

CHAPTER 5

Overgrazing

FOR DECADES "OVERGRAZING" WAS THE ENVIRONMENTAL CRITICISM most commonly lodged at cattle and beef. Implicit in the word is the idea livestock are responsible for much of the globe's denuded landscapes and desertification. The term also suggests the problem boils down to far too many animals eating too much vegetation and doing too much trampling. Books like Jeremy Rifkin's widely read *Beyond Beef* urged people to stop eating beef almost entirely based on the argument that the American West and much of the globe has long suffered from serious overgrazing. Rifkin portrays an American cattle rancher despoiling the natural environment, unregulated, government-sanctioned, and feeding at the public trough.

Soil erosion and desertification are very real. Just as I believe in human-caused climate change, I have no doubt erosion and desertification are happening and have been largely caused by human activities. Globally, erosion is now occurring more than 20 times faster than the geological rate. In some areas the situation is much worse.[1] "The estimated rate of world soil erosion now exceeds new soil production by as much as 23 billion tons per year." Left unabated at this pace, the world will run out of topsoil in little more than a century.[2] Meanwhile, "The [worldwide] rate of desertification is estimated at 5.8 million hectares (Mha) per year," notes one credible source.[3]

Yet assertions that these problems result from too much grazing—historically and today—are usually unscientific. The real question is not whether these phenomena are occurring, but what is actually causing them. Equally important, what can humans do to stop or even reverse the trends as we feed ourselves now and in the future?

Looking back at the history of once fertile areas that are now deserts, it's routinely stated (with scant real evidence) that overgrazing caused the transition. Regarding the Middle East, Steven A. Rosen, professor in the archaeological division at Ben-Gurion University, Israel, says, "It is a common conception that desertification and the destruction of previously fertile and productive lands are often the result of pastoral nomad activities, in particular through overgrazing by goat and sheep herds." The assumption, however, does not withstand scrutiny. Rosen notes that a critical analysis, using archaeology as the primary tool of historical reconstruction, showed "no evidence for causal links between pastoral nomadism and desertification." Additionally, the historical reconstruction analysis "suggest[s] that declines in pastoral nomadic presence in the desert are related primarily to changes in the sedentary infrastructures with which the nomadic systems interacted."[4] In other words, human development of the landscape, rather than pastoralism, was to blame.

Grazing often gets wrongly lumped in with damage from other human activities. Usually, it's crop cultivation. Professor David Montgomery's book *Dirt: The Erosion of Civilizations* documents a worldwide pattern: poorly managed crop farming leading to soil exhaustion and erosion, followed by grazing animals. Then people blame livestock for the land's poor condition. Montgomery's book chronicles examples of such mistaken observations of 19th-century travelers. They attributed dramatic population shifts in Jordan and Tunisia to soil losses they wrongly believed had been caused by overgrazing.[5] In fact, as has happened elsewhere, the land had first been cultivated and depleted to the point that it would no longer support crops. Livestock were simply the final, default use in a long history of poor land stewardship.

Montgomery's review of human societies around the globe depicts a typical scenario. Civilization after civilization settled in rich, fertile valleys and grew crops. Then, as populations expanded, they pushed crop farming up into the hills, often cutting down vast forests. But slopes, especially after being clear-cut and farmed in the same manner as valleys, quickly lose their virgin topsoils. Crop cultivation, we know, leaves part of the soil bare some or all of the time. This makes it especially "vulnerable to erosion, whether by wind or rain, resulting in a rate of soil loss tens to hundreds of times faster than nature makes it," notes Montgomery.[6] Plants and the litter they produce protect the ground from the impact of raindrops and action of flowing water. But bare ground is exposed to wind and rain. The blast from each incoming raindrop sends dirt downslope.[7]

Over time, societies who followed this path lost their capacity to feed themselves. Much of the drive for Roman conquests, Montgomery convincingly argues, was fueled by poor agricultural practices whittling away the productivity of the empire's cultivated areas. Montgomery suggests exhaustion and erosion of the soil was a major factor in the fall of most once thriving civilizations, including the ancient Greeks, Romans, and Mayans. "[T]he human cost of soil exhaustion," he writes, "is readily apparent in the history of regions that long ago committed ecological suicide."[8]

There can be little doubt that since its beginnings some 12,000 years ago, agriculture has done inestimable damage to the globe. "Most analyses of problems *in* agriculture do not deal with the problem *of* agriculture," begins Wes Jackson's classic treatise *New Roots for Agriculture*. "[T]he plowshare may well have destroyed more options for future generations than the sword."[9] A bit later he continues: "So destructive has the agricultural revolution been that, geologically speaking, it surely stands as the most significant and explosive event to appear on the face of the earth, changing the earth even faster than did the origin of life."[10]

Speaking generally and of the United States, Wes Jackson also writes: "The hills are living, and, so long as they are clothed, eternal; the relatively flat lowlands, put to the plow by scarcely three generations of land stewards, are ephemeral."[11] American settlers followed the nearly universal pattern of failing to carry out sound agricultural land management practices, especially the use of cover crops and the application of animal manure, which are necessary to avoid erosion and keep soils living and fertile.

While stating hopefully that "[g]ood management can improve agricultural soils just as surely as bad management can destroy them,"[12] David Montgomery notes "[d]ifferent types of conventional cropping systems result in soil erosion many times faster than under grass or forest."[13] The problem, he explains, is that soil organic matter declines under continuous cultivation as it oxidizes when exposed to air. When soils have more organic matter they have more glomalin, and more soil aggregates, all of which inhibit erosion. "Thus, because high organic matter content can as much as double erosion resistance, soils generally become more erodible the longer they are plowed."[14]

It has often been suggested the past missteps of agriculture could be corrected by reducing or removing animals—especially cattle and other grazing animals—from the world's food system. But as the previous discussions on carbon sequestration, soil, water, and biodiversity have illustrated, that would be moving in precisely the wrong direction. Humanity's greatest agricultural misdeeds have not been carried out with grazing animals. It has been the plow. And our greatest crime against nature has been ripping asunder the dense plant cover that once protected our earth.

Grazing animals' greatest asset is they can be raised for food without any plowing. They can live on and be fed by grass. This benefits the world's climate and provides essential ecosystems for a menagerie of beasts from the tiny to the mega. Even poorly managed rangeland is ecologically preferable to cropland. "Many ranchers do overgraze and their soil does erode," Jackson writes, "but even with overgrazing, poor ranching and no ethic, the land fares generally better when grazed than when put to the plow."[15]

I certainly agree livestock's impact on the earth has not always been positive. Here, the historical American West is a good case in point. Geography professor Paul Starrs, in his book *Let the Cowboy Ride*, argues that American ranches have been culturally and ecologically important. Still, Starrs notes, the forest reserve system was established in the early 1900s as a response to "virtually universal overgrazing in the Sierra Nevada of California and the mountains of northern New Mexico."[16] He adds, "The years from 1880s to the mid-1930s marked a half-century of destructive exploitation of western rangeland that has few North American equivalents for intensity of rapine or unregulated voraciousness."[17]

However, this picture of cattle grazing is increasingly out of date. As Starrs points out, US grazing has long been much more carefully managed than it was at the turn of the century. He attributes this to several factors, including governmental regulations (starting with the 1934 Taylor Grazing Act, and continuing with environmental laws adopted in the 1970s) and a newly emerging scientific understanding of grazing's complex ecological impacts. Just over the past decade, I have witnessed a major upswing of interest among ranchers in improving both animal husbandry and grazing practices.

Recent years have also seen cattle numbers come down, and there is much less grazing on public lands. Popular perception has long been that the United States has ever-rising numbers of cattle, especially on public lands. In fact, the reverse is true. The federal Bureau of Land Management, which controls most grazing on public lands, reports that from 1954 to 2013 grazing on public lands declined by a whopping 57 percent.[18]

A lot of people have heard, or believe they have seen for themselves, how cattle have damaged America's West. Yet, like The Nature Conservancy's vernal pool studies, field studies measuring actual impacts have surprised cattle's critics. One investigation of grazing's impacts by the USDA Forest Service found removal of grazing in riparian areas might be beneficial for the short term, but would be "neutral to negative" over the long term. In part, the findings read:

> [G]rasses have evolved with the periodic removal of vegetative material through fire, insects, or ungulates. In the absence of grazing or other disturbance, plants continue to accumulate litter (dead grass blades left at the end of the growing season). After years of litter accumulation, plants go into a self-imposed stress whereby the detritus (previous years' growth) chokes out new shoots competing for light. The vigor of the entire plant is compromised and rangelands become less productive and healthy. Many invertebrate and wildlife species depend upon productive grasslands, especially for winter range. In addition to loss of plant vigor and decrease in rangeland health, the accumulation of litter allows fine fuels to build, which increases susceptibility to fire.[19]

Even more important, I believe it's a misnomer to label historical damage to America's West overgrazing. Use of the word confuses the issue. It fosters fundamental misunderstanding. *Overgrazing* is a word fully loaded with the idea of excess. Excesses of animals, of eating, of pooping, and of trampling. Cattle can cause environmental damage to the land. But no one has proven damage to the American West from *too many* livestock. I believe the land has not been overgrazed. It has been *improperly* grazed. The distinction is crucial.

Why do I say this? Because of history and evolution. For millions of years, the earth was covered with vast herds of large grazing animals. The 70 million bison extant in 1800 were just a remnant of the populations of large herbivores once blanketing the North American continent. Pressure from abundant predators kept grazing animals in close, dense, constantly moving formations. The world's remaining large wild herds—like caribou in North America and wildebeest on the Serengeti—still function this way today. The massive herd rolls along the landscape as a single, giant organism—eating, pooping, and trampling. Then the herd moves off and does not return for an extended period of time. It was under such conditions—intensive animal impact—that most of today's land-living organisms and ecosystems evolved.

Present-day restoration efforts are also instructive. We know that where similar conditions have been re-created in the modern world, dramatic improvements to land, water, and biodiversity have resulted. If we only allow our anti-cattle blinders to fall from our eyes for a moment, we can clearly see the logic, and the biological, historical, and real-world evidence. From this evidence we know it is not cattle, or grazing, or even overgrazing, but the way that cattle are managed that is the problem. Again, I say: "It's not the *cow*, it's the *how*."

Simply putting grazing animals out on the land won't trigger the full ecological benefits. Today's world is woefully understocked in predators, an ecosystem element that causes animals to bunch densely and move frequently. Thus, management techniques must be adopted to keep animals together and on the move, mimicking the movements of earlier wild herds. Attentive stewardship of livestock is imperative. The earth needs grazing but it needs well-managed grazing, and lots of it.

Fortunately, some reconsideration of overgrazing is occurring—even in mainstream ecological discussions. A 1998 report for the World Bank and United Nations was a harbinger of some of this shifting understanding:

> *Conventional wisdom suggests that much of the blame for "desertification" and land degradation in arid rangelands rests with pastoral livestock production. There is now considerable literature which corrects this misconception on two counts: the extent of dryland degradation is greatly exaggerated because underlying ecological dynamics have been misunderstood, and the contributory role of livestock has been mis-specified.* [20]

The newer and better understanding has included increased attention to important factors like grazing intensity and timing. "While literature from the 1980s and early 1990s repeatedly linked livestock to the degradation of rangelands, more recent studies have refuted this by suggesting that prolonged rest leads to even more serious degradation," states a 2011 article in the journal *Pastoralism*.[21]

The major study of soil by Lal and Stewart, discussed earlier, notes, "Desertification is a biophysical process (soil, climate and vegetation) driven by socio-economic and political factors." Looking toward solutions, their paper urges reversing desertification would "improve soil quality, increase the pool of carbon in soil and biomass, and induce formation of secondary carbonates leading to a reduction of carbon emissions to the atmosphere." To do this, it first recommends "establishing vegetative cover with appropriate species."[22] Logically, this would include both climate-appropriate grasses and the animals needed to maintain them.

Evidence is also mounting that livestock are essential to global environmental and food system solutions. In addition to the benefits of grasslands and grass areas on farms, animal manure is being recognized as an asset the world food system cannot afford to be without. We've seen how adding manure to soil helps fertility and reduces erosion. A University of Washington study made direct measurements of soil losses between 1948 and 1985. It found that a chemically fertilized field lost four times as much topsoil as an organically farmed field, fertilized only with manure.[23]

Hooves, mouths, and digestive tracts are also directly repairing damaged land. In what's being called a "cattle stomp," ranchers and land preservation groups in Colorado are working together for long-term land regeneration on a former mining operation. Integrating biochar and compost and using rotational grazing, the stomp consists of native grass seed covered by a layer of wheat straw, where cattle are fed hay. The acreage is fenced off into small plots. As cattle graze hay in each plot, their hoof action "stomps" grass seed and straw into the soil, and their waste products provide moisture and natural fertilizer.[24]

From the historical to the current, real-world evidence shows cattle grazing and trampling to be beneficial, even essential. In those situations where cattle and other grazing animals have caused environmental damage, the problem is one of improper management rather than overgrazing. For grazing to function as it should, it must be planned, and cattle must be actively managed. With such human involvement, cattle fulfill the essential ecosystem role of grazing, manuring, trampling herbivore.

CHAPTER 6

People

THIS BOOK ARGUES BEEF IS HEALTHY FOOD, AND CATTLE, AS herbivores that create biologically vibrant soils, and convert sunlight, captured in the world's grasses, to meat and milk, are essential to a viable food system. Much has been said, and will continue to be said, on both sides about these hotly contested issues.

Other benefits of keeping cattle are less obvious and barely part of today's public dialogue. Here's one I've thought about a lot: We need good jobs for our citizens and we need *people* in agriculture. Cattle enable people to make a living doing it. Both varieties of bovines, dairy and beef, add value to diversified farms. Beef cattle also uniquely make possible, especially in arid and semi-arid areas, like America's Far West, the existence of homesteads and accompanying enterprises on remote non-tillable land, where crop farming is environmentally inadvisable or impossible.

Being a professional farmer or rancher has never been easy. It's particularly challenging these days. The US Department of Agriculture reports more than 93 percent of people involved in American agriculture now have at least one other income source. Most people who actually work their own lands make, at best, a modest living. With such slim margins, animals are often invaluable.

I've met farmers raising diverse crops—cereals, vegetables, fruits, and livestock—who've told me their farm's survival is owed to their animals. Livestock and poultry are often the most challenging yet most profitable aspect of a farming operation. Many of the ranches I know (including our own) are on dry, windy, or hilly lands where rain falls sparsely or only during the coldest months of the year. For both environmental and economic reasons, these lands cannot be used to grow crops. Other stretches of the globe, including in the United States, have been so eroded and exhausted by plowing and crop cultivation they can now support *only* grazing animals. All these physical conditions can support grasses, and the animals that thrive on them, but not crop farming. Simply put, cattle often are what makes a living in agriculture financially viable.

Some may ask why this matters. These days, only 17 percent of Americans reside in rural areas and fewer than 1 percent are directly involved in agriculture. As I write this, the COVID-19 pandemic is reshaping Americans' views on the desirability of city dwelling. Yet many still consider urbanization inevitable and the disappearance of family-scale farms and ranches unavoidable. But I strongly protest.

For starters, we need more, not less, of what we eat to come from true family farms and ranches. Food from known, trusted, transparent sources. Food that supports our local economies. Food more likely to be fresh, organic, and humanely and regeneratively produced. Food that's not genetically modified.

Independent family farms are the ones preserving the varied strains of domesticated plants and animals. Hearty breeds of cattle, poultry, fruits, vegetables, pulses, and grains that generate delicious, nourishing foods. The COVID-19 global pandemic is waking people up to the failings of an industrialized food system. It lacks genetic diversity, produces vast monocultures, and depends on global rather than local markets. Its fragilities have now been laid bare. Locally produced foods have proven less vulnerable to supply chain interruptions. Many now appreciate, for the first time, the value of "old-fashioned" genetic diversity that resists novel diseases.

Real farms and ranches, especially those with animals, also create an incomparable living environment for *humans.* Over my life, starting in early childhood, I've visited scores of farms and ranches and met hundreds of people making their living in agriculture. I've seen how working the land forms unique individuals. Farmers and ranchers are resourceful, handy, hardy, courageous, independent, unflappable.

Childhood on a farm or ranch is formative. It makes strong bodies and builds character. Daily, physical, outdoor labor in rain, snow, wind and heat engenders physical toughness and stoicism. Being dependent on the vagaries of nature—where months of work can be destroyed by a single fire or storm—teaches humility and resilience. Being far from town means there is usually no one to call to come fix things. Farmers and ranchers are, by necessity, mechanics, plumbers, and builders. All of this fosters creativity, handiness, and, most of all, self-reliance.

I've also found farm and ranch people infused with a deep and true environmental ethic. They depend on weather and seasons, staying closely attuned to the events of the natural world. They're immersed in the cycles and rhythms of life, intensely focused on the health and well-being of the plants and animals surrounding them. Courtship, mating, birthing, caring for young, growing, injury and illness, aging, and, perhaps most important, dying—form the daily experience.

And because working the land and caring for animals creates people connected with nature, it makes some of the best environmental stewards. A country benefits by having a good portion of its population in agriculture and connected to the natural world. It's a connection worth preserving. As a nation, something intangible, yet vital, is slipping away as the number of farming and ranching families dwindles. It is diminishing our national character.

Farm life is not akin to living in the country while telecommuting to an urban job. The farm environment does not exist without farms. And it is a human environment we should protect.

The hard work of farmers and ranchers is also unique because it's practiced by whole families, together. On many of the ranches I know, 3-year-olds have daily chores and 5-year-olds are on horseback; 13-year-olds drive trucks while 15-year-olds operate tractors. Children are given important tasks with real responsibility. They contribute meaningfully to the family's collective job, even their capacity to make a living. And these children rise to the occasion. The farm kids I meet continually impress me as responsible, even-keeled, and well spoken.

These children labor, yes, but they also play, and the play is grand. The great outdoors is their oyster. Like modern-day Huck Finns and Laura Ingallses, nothing stands between them and countless real adventures. Their days aren't filled with outings contrived by overseeing parents or teachers. Farm and ranch kids spend hours, daily, exploring the natural world with minimal interference from adults. Through their own direct experiences, they develop maturity, knowledge, and curiosity about nature.

At one ranch Bill and I visited, the eight-year-old son, after politely greeting us and chatting articulately for several minutes, excused himself with his fishing pole and sack lunch in his hand, his dog by his side. He was going, he explained, to his favorite fishing hole. He would be back before dinner. It was the first time since my youth I'd seen a child under 10 with that much independence.

When I was a kid, the gap between farm and suburban life was not the chasm it is today. Growing up in the outskirts of Kalamazoo, Michigan, in the 1970s and '80s, we were regularly on farms and had daily chances to play and explore outside near our home. At least a dozen times a year my mother and father took my brother, sisters, and me to farms west of town. We U-picked fruits, bought eggs, vegetables, and cider directly from farmers. In December, we painstakingly selected and sawed down our own Christmas tree.

My daily dose of nature came closer to home. Steps from our family's door was an open 200-acre parcel, a mixture of grassy, scrubby, and wooded land. To the neighborhood's adults, it was just a nondescript, undeveloped parcel or a place to walk to the dog. But for us kids, it was a vast, mysterious wilderness rife with possibilities. Here we explored and learned on our own. We spent time with trees, wildflowers, fungi, insects, rodents, birds, toads, foxes, and deer. Like most kids of that era, I had broad freedom to roam. Part of nearly every day was passed in this undeveloped acreage, climbing trees, foraging berries, turning over rocks, following animal tracks, gathering flowers, seeds, leaves, and pinecones, and simply looking and listening. I remember lying alone on my back, watching clouds float past. Sometimes I'd bring a book to read. Other times, I'd carry a backpack with a snack, a sketchbook, and some pencils. On Sundays, our entire family would often go there for a long stroll, then return home to warm up around the fireplace.

This seemingly undistinguished patch of land, bursting with life and constantly changing with the seasons, was my second home. I told my mother I was most certain there was a God when I was out there among the trees.

In college, majoring in biology, I studied cellular metabolism, ecosystems, and how plants and animals develop and grow. I relished the fieldwork. To gather plant and insect specimens, I returned to the plot of land I knew so well. And for years to come, whenever visiting my parents, I walked those fields, usually with my father and his faithful canine companions. That same plot of land, when later threatened with commercial development, drew me from North Carolina back to my hometown.

Growing up outside and deeply connected with a particular wild place, the way I did, is a rare experience for children these days. In today's urbanized and suburbanized environments, children spend hours in front of screens, far less time outside, and little time in nature. Nearly every outdoor experience happens under adult supervision. We think of this as normal, and rarely reflect on the consequences of such an altered way of rearing our young. Kids who grow up on farms and ranches are the exceptions to this norm.

In his book *Last Child in the Woods: Saving Our Children from Nature-Deficit Disorder*,[1] journalist Richard Louv argues strongly and convincingly that youth need time in nature. Children themselves, and ultimately our country as a whole, he urges, are injured when children are divorced from nature, as so many are today. "If children do not attach to the land, they will not reap the psychological and spiritual benefits they can glean from nature, nor will they feel a long-term commitment to the environment, to the place,"[2] Louv writes.

Occasional forays into parks and watching nature programs won't do the trick, Louv urges. "Passion does not arrive on videotape or on a CD; passion is personal. Passion is lifted from the earth itself by the muddy hands of the young; it travels along grass-stained sleeves to the heart."[3]

Living on our ranch for the past seventeen years, Louv's words ring true to me. When I moved here after nearly five years in New York City, I felt like a pig wallowing in mud on a hot summer's day. I experienced a deep and visceral euphoria and peace at returning to life amid the natural world, the same joy and calm I'd found as a child. And I understood fully, for the first time, how much I needed that time in nature to feel physically, mentally, and spiritually healthy and whole.

My experiences have also given me perspective on the modern divide between urban and farm life. I've seen a widespread disconnectedness from nature. We frequently host visitors here, usually chefs, environmental or animal activists, retailers, journalists, friends, or family. The degree to which they—adults and children alike—are unfamiliar with nature's workings often becomes startlingly apparent.

I remember a young woman visiting from Washington, DC, who was shocked and upset at seeing a nanny goat butting away a kid goat who tried to nurse from her. This was not a vicious butting. It was the rather typical *Get away from my udder* butt you might see any day in any group of goats or cattle. "Oh, that's normal. That's not her kid," I explained. When this failed to soothe our distressed visitor, I offered further explanation. "The mother needs to protect her milk for her own baby," I said, "and that kid has a mother of its own that he needs to go find." Nothing I said changed the opinion of our visitor. To her, that was just one mean goat.

I recall a famous young chef from New York City. He blanched and looked like he might be ill when our Great Dane, Claire, bounded up to us proudly displaying a cherished prize in her mouth: a whole deer leg left over from a coyote feast.

I can't forget an alert, happy, almost five-year-old boy visiting from San Francisco. He was unfamiliar with the most basic biological facts. As we walked through our pasture surrounded by turkeys I began by speaking of the "males" and the "females." When I saw that wasn't resonating, I tried calling them the "boy turkey" and "girl turkey." Still nothing. Finally, I looked to the child's mother for help. "Maybe if you said 'mama turkey' and 'papa turkey,'" she suggested. When that, too, failed to be understood, I finally realized the boy did not know animals come in two distinct genders.

The implications of our modern disconnectedness from nature and animals are manifold. I agree wholeheartedly with Richard Louv's suggestion it broadly affects attitudes. "To urbanized people, the source of food and the reality of nature are becoming more abstract. At the same time, urban folks are more likely to feel protective toward animals—or to fear them."[4]

Beyond this, I am increasingly persuaded this disconnect is even making modern humans less comfortable with the inevitable events of our own lives. We will experience aging and physical decline. We will live through sickness, injury, pain, and, ultimately, death. Experiencing all of this through the lives of other living organisms gives us perspective that we are part of a much larger cycle of life. It prepares each of us for our own journey.

Life on a ranch lends itself especially well to gaining this understanding. Kids see, touch, hear, and smell plants and animals every day. Plants are sprouting, growing, blooming, fruiting, drying, dropping seeds, dying. Animals—wild and domesticated—are mating, being born, nursing, growing, fighting, getting sick and hurt, and dying. Owls, ravens, hawks, snakes, spiders, coyotes, bobcats, gophers, rats; cattle, horses, dogs, cats, chickens, turkeys, goats, donkeys, mules. Feathers, fur, bones, dust. These are the ingredients of our days.

Every day involves hearing and observing (consciously or not) wild animals. And much of the day will be spent caring for (willfully or not) farm animals. We provide feed and water, check on them, search for them, doctor them, and, more likely than not, we talk and sing to them. The needs, wants, and habits of creatures become intimately familiar to all who live alongside them. Thinking beyond yourself becomes second nature.

Years ago, before I married Bill, a young couple named Rob and Michelle Stokes lived here and ably carried out the day-to-day management of our land and our cattle. Now they have a place of their own in Oregon, and two teenage daughters. They raise cattle, pigs, goats, and heritage turkeys, all on grass. At one point, they spent several months back on our ranch and managing a large goat herd here, along with their first daughter, just three years old at the time.

One day Michelle asked for my help with a mother goat she'd been treating for pneumonia. Initially, the goat seemed to have responded well to treatment. But that day she'd unexpectedly taken a turn for the worse. She was now out in a pasture and too weak to stand. Together Michelle and I brought her back to the barn and laid her on some hay, out of the wind. We were trying to revive her with a liquid nutrient supplement. "C'mon, girl," Michelle kept softly urging her. She died a while later with her head in Michelle's lap. Michelle's daughter stood a few feet away, watching and asking questions. As she doctored the goat, Michelle unwaveringly explained to her daughter what was happening, gently yet forthrightly answering her daughter's inquiries. I was deeply moved by the kind and skillful way Michelle handled her child's curiosity and concern.

Years after that event, Michelle and I communicated about the value of raising children on a ranch, and it reminded me of the day we had tried to save that nanny goat. She wrote:

My rancher children, in their short five and nine years, have seen more loss of life than most individuals in a life time and yet we still mourn these losses and understand loss. They are also blessed to be involved in the miracle of birth and the blessings of life over and over. They are witnesses to the struggles of saving and losing animals we love and they realize that there are many times that other lives take precedence over our own wants and sometimes our needs, truly giving them empathy for all living creatures.

Now that Bill and I have two children of our own, I, too, see, every day how much they gain from such a life. A few days after his entrance into the world, our older son Miles began going out to do chores nearly every day with Bill or me. By two years, most days he played for hours out in the yard. Out on the ranch, he comfortably ambled among turkeys that towered over him and eagerly begged to be useful.

We began giving him small, albeit real, jobs like gathering turkey eggs with us, helping to herd the turkey flock (but not yet the cattle!), and filling waterers. By age five, he was opening and closing gates, rounding up stray turkeys, and learning about nearly everything. Now, at eleven, he drives the farm truck, helps fix whatever breaks, and provides indispensable assistance in countless other ways. His help is invaluable to us, and the pride he feels in his contributions is evident.

Our younger son, Nicholas, age seven, has always been an eager and able helper. He loves to go out with his papa and feed hay, gather eggs, or check on the cattle. When he returns to the house, he often finds me in the kitchen and recites in tremendous detail every way he helped out.

Like other ranch kids, our boys spend hours outside every day, running, climbing, breathing fresh air, soaking up the sun. They are surrounded by animals, of which they feel no fear. (We have to teach them when they *should* be afraid.) Both boys can identify dozens of local birds, insects, flowers, and trees. They are energetic, strong, and rarely even notice their skinned knees, scratched legs, and dirt-stained clothes. Summer days are spent outside from dawn till dusk.

In their remarkable book *The Geography of Childhood: Why Children Need Wild Places*, Gary Paul Nabhan and Stephen Trimble describe the connection between children and nature. Children should be reared so they spend much more than occasional, carefully supervised time in natural spaces, they urge. Comparing their experiences to those of the Hopi in Arizona, they write about children reared on a ranch:

> *Ranch children learn in similar fashion about careful observation and nuance of behavior by taking care of horses and pets and stock. They also understand domesticated animals as utilitarian, existing to serve human needs. Death may move them, but it does not shock them. Their intimacy with animals is an intimacy with life.*[5]

On our ranch, as on every farm and ranch with cattle, understanding nature is fundamental to the family's craft. In addition to paying attention to her animals, a good rancher knows the area's patterns of rainfall, wind, and temperatures. She grasps how they affect land, vegetation, and animals. She is knowledgeable about local wild flora and fauna; she is familiar with the area's rivers and streams and how they function. In short, a good rancher is not merely working outside. She is constantly interacting with—*depending on*—the natural elements in which she is functioning. It's the ideal incubator for naturalists.

And we need naturalists now more than ever. Scientists study nature, but not all of them are true *naturalists*. These are people who love and understand the natural world in the tradition of Charles Darwin, Beatrix Potter, Henry David Thoreau, Aldo Leopold, and Wendell Berry. Dr. Paul Dayton, professor of oceanography at Scripps Institution of Oceanography in La Jolla, California, points out that at the very moment when the earth is under its greatest assault, academe virtually eliminated direct natural experience from its curriculum. This, Dayton argues, takes away "the opportunity for both young scientists and the general public to learn the fundamentals that help us predict population levels and the responses by complex systems to environmental variation."[6]

Our planet's best hope is having current and future generations who truly understand and cherish nature. "If we are going to save environmentalism and the environment," Richard Louv writes, "we must also save an endangered indicator species: the child in nature."[7]

Farms, especially those with animals, have long been among the best places for children to gain deep understanding of biology and to fall in love with nature and all its plant and animal components. Louv tells the following story of a longtime environmental activist:

> *When Janet looks back on her childhood in nature, she sees it not only as the source of her environmental activism—her work protecting the mountaintops of West Virginia—but also as nourishment for her own spirit. Her favorite place was a dairy farm run by her aunt and uncle . . . Off she would dash—to the barn, the henhouse, a hillside, meadow, or creek to explore the rich, natural treasure trove that lay before her. Whether she was watching the birth of newborn kittens or mourning the loss of a baby bird found featherless and cold on the ground, nature provided Janet with ample opportunity to feed her curiosity about life and taught her about the inevitability of death.[8]*

Besides being future environmentalists, people reared wholly or partially on farms and ranches become especially productive members of society. Before I started working as an environmental lawyer, I was only occasionally in the presence of cattle or the people who raised them. But I had been to quite a few small, diversified farms in southwestern Michigan. I'd seen enough to understand why Cato and Thomas Jefferson both exalted the farmer as the ideal citizen. The farmers I remember from my youth were hardworking, doggedly independent, and civic-minded. They loved the land and had deep firsthand knowledge of the local climate and soils. They knew the area's flora and fauna, and were banks of local history lore. Of course they would be valued in society.

As Waterkeeper's senior attorney, I began working with farmers and ranchers from around the country, and spending more time on farms. I was again struck by the uniqueness of this set of people. And I began to see how the decline in the number of people in agriculture was more than just a national employment concern. It's draining a pool of uniquely valuable citizens.

This is not just my own wild idea. This point was echoed by a prominent CEO of a Fortune 500 company being interviewed on the radio. The interviewer asked what the executive looked for in a job candidate. I expected his response to include something about universities or internships. But without hesitation he replied his first choice was always a person who grew up on a farm. Years of experience hiring and supervising people, he said, had taught him farm kids were dependable, hardworking, and resourceful.

Later, I saw a talk by a retired military officer who'd started a farm with his wife. Although he hadn't grown up on a farm himself, he said he wanted to raise his children on one. Decades in the military had taught him the best service people were those raised on farms. They were polite, disciplined, and reliable. This, he said, was just how he wanted his kids to turn out.

I suspect many college admissions directors would view things similarly. My parents, who spent their careers teaching at universities, both said farm life formed some of their most diligent students.

Statistics showing the disappearance of farmers and ranchers are often recited and lamented. Yet one aspect of America's shifting demographic has been largely ignored: As we lose farms and ranches, we are also losing an exceptional environment for child-rearing and an incubator for good citizens.

Equally underappreciated is that we are losing a physical environment conducive to rearing strong, resilient children who become healthy adults.

As our son Miles was still in the womb, I did several hours of chores on the ranch, every day. Clearly, I had no choice but to bring Miles with me. (Well known among friends and family is the infamous story of Miles's birth: My water broke while we were out on the ranch during morning chores, followed by me attending Easter Sunday services and singing a Bach solo before heading to the hospital.)

Occasionally, I wondered if it was such a good idea to expose my unborn child to the dust from turkey feed, hay, manure, and dander from cattle and turkeys. I should not have worried. New research is turning my concern on its head. Rather than harming my sons, my working around animals has likely been protective of their health, as will be their experiences growing up with livestock.

A 2013 *New York Times* op-ed explains why many scientists now consider children who grow up on farms with livestock to be the best protected from allergies and asthma.[9] Part of the research was triggered by the discovery that the Amish, 92 percent of whom often spend time on farms, have extremely low rates of allergy and asthma. "Westernization," or maybe more accurately "industrialization" (which the Amish have strongly resisted), has been connected with people becoming allergic. Until fairly recently, it was not understood why.

Now it is believed that the so-called farm effect protects farm kids by stimulating their immune systems. Humans on farms and ranches, beginning even before birth, are chronically exposed to high levels of microbes, such as from cattle manure, dander, and hay. The earlier the exposure starts, including during pregnancy, the greater the protection. "Children born to mothers who work with livestock while pregnant, and who lug their newborns along during chores, seem the most invulnerable to allergic disease later," the article notes. Research by Bianca Schaub, a doctor and researcher at Munich University, shows newborns from farms are better at quashing allergic-type reactions. "The more cows, pigs and chickens a mother encounters, essentially, the more easily her offspring may tolerate dust mites and tree pollens."[10]

The research also suggests that mothers living on livestock farms and ranches might provide a natural therapy to the baby's immune system, "one that pre-programs the developing fetus against allergic disease."[11] Increasingly, science supports the conclusion recent decades' epidemic levels of allergy and asthma are linked to our modern-day lifestyles, especially our disconnect with farms and farm *animals*.

Dr. Maya Shetreat-Klein is a medical doctor, mother, and author of the book *The Dirt Cure*. She argues the reason farm kids are so vigorous is not so much about the quantity of microbes they are exposed to but the diversity. "Microbial diversity seems to have a very powerful impact," she says. The more types of microbes children encounter, "the more likely they are to be in balance, and the less likely they are to let any one micro-organism grow out of control, as occurs with infection." She draws a parallel with plant health. "Phytonutrients are part of the plant's immune systems," Dr. Shetreat-Klein explains, and pests stimulate plants to make more phytonutrients. "Being exposed to different organisms improves the health of the plant and it improves our health as well."[12]

The importance of exposure to germs is actually not a new idea. Jared Diamond's brilliant book *Guns, Germs, and Steel: The Fates of Human Societies*, first published in 2006, includes the argument that historical events demonstrate the importance of exposure to livestock, especially cattle. Dissecting the complexity of reasons certain cultures became more technologically advanced and, consequently, more politically and militarily dominant, he argues that domesticated animals, especially cattle, helped humans develop immunities to infectious diseases.

By geographic accident, some cultures arose in regions where animals existed that were prone to domestication. Cattle were particularly significant because they could pull carts and plows, as well as provide meat, milk, and manure. Living in close proximity to domesticated animals, humans exposed themselves to more disease-carrying organisms. On the one hand, this contributed to epidemics. On the other hand, Diamond argues, coming in contact with the germs carried by cattle and other animals was critical to the successes of those cultures. Being exposed to the diseases led to these populations developing immunities.

However, when those societies collided with cultures that had no domesticated animals, and therefore had not been exposed to livestock-borne diseases, the results were catastrophic for the cultures without livestock. The vast majority of the deaths wrought by Europeans in the New World were from their germs. These were pathogens to which the native peoples of the Americas had never previously been exposed. The spread of infectious diseases massively abetted the military defeats inflicted upon Native American cultures throughout the American continents.

In addition to the immunological benefits of being around lots of germs, the farm kids I know (including our own) tend to spend much more time outdoors and much less time in front of TVs and computers than most kids. The typical American child now spends an average of three hours a day in front of a television and five to seven hours a day in front of a screen (which includes video games, looking at Facebook or Twitter, and otherwise using a computer).[13]

Although most kids start the habit early, the American Academy of Pediatrics recommends children two and under have *zero* "screen time." Allowing young kids too much time in front of screens has been linked to various behavior problems, especially attention deficit hyperactivity disorder (ADHD).[14] The increased amount of time children spend staring at screens has also been linked to the rise in obesity and vitamin D deficiency.[15]

"Several large-scale studies have found that vitamin D deficiency is widespread—one in ten U.S. children are estimated to be deficient—and that 60 percent of children may have suboptimal levels of vitamin D," states the website of the Johns Hopkins Children's Hospital.[16] Normally, 90 percent of a human's vitamin D comes from time spent in sunlight.[17] Getting sufficient vitamin D from foods is difficult, and research suggests vitamin D pills are not equivalent to sunshine. It's yet another reason being outside every day is vitally important to good physical and mental well-being.

These issues—the physical environment and quality of ranch life, especially for children—may seem only remotely related to cattle. But they are real and important. Cattle enable the existence of many American farms and ranches. If cattle disappeared from our food system, these human environments would disappear as well. This would be a tremendous loss to our culture and society. It's a loss we cannot afford and should not abide.

CHAPTER 7

Health Claims Against Beef

The Rest of the Story

FEW DOUBT WE ARE IN THE MIDDLE OF A PUBLIC HEALTH CRISIS connected with the American diet. Clear and shocking statistics show a country whose people are getting fatter, have higher blood pressure, and are increasingly suffering and dying from chronic diseases, including heart disease, diabetes, and stroke.

At the center of the storm is our widening girth. Two-thirds of American adults (69 percent) are now overweight, with 39 percent obese (men having more than 25 percent body fat, and women with more than 30 percent). From 1960 to today, the average weight of American men rose from 166 pounds to 191 pounds; women's average went from 140 pounds to 164. So many human health problems are triggered by carrying too much body weight that the dramatic rise in obesity is cause for great concern.

For young people, the situation is graver still. When I think back to my childhood in the 1970s, I can remember just one child, in a neighborhood chock-full of kids, who would meet the definition of obese. However, from 1970 to 2000, the number of overweight children more than doubled, and the obesity rate tripled. Nearly one-third (32 percent) are now overweight, with 17 percent obese. The American Public Health Association recently averred that the situation is "leading to a generation at risk for cardiovascular diseases, diabetes and other serious health problems."[1] Dr. Robert Lustig, a physician at University of California–San Francisco who has clinically treated and studied obese children for the past two decades, reports it is even affecting babies. "We now have an epidemic of obese six-month-olds," Lustig warns.[2] (Some of the reasons why babies are affected will be discussed later.)

Closely related to the trend of heavier weights is the rise in chronic conditions and diseases. One-third of US adults (33 percent) now have hypertension. Four of five leading causes of US deaths are chronic diseases.[3] In order, they are: heart disease, cancer, accidents, respiratory disease, and stroke.

The trends are particularly unsettling. Obesity, hypertension, and chronic diseases are increasing or merely leveling off despite several factors that *should* have dramatically reduced them. Take heart disease and its related deaths. Advanced medical interventions for heart attacks are available at every hospital; stent insertions and coronary bypass surgeries have become routine. Authors of a 2011 *Journal of the American Medical Association* article on heart procedures noted: "Coronary revascularization, comprising coronary artery bypass graft (CABG) surgery and PCI [stent insertion], is among the most common major medical procedures provided by the U.S. health care system, with more than 1 million procedures performed annually."[4] At the same time, smoking, among the top risk factors for heart disease, is much less common. Adult male smoking rates plunged from 55 percent in the 1950s to 18 percent today.[5] Nonetheless, heart disease incidents and death have merely leveled off in the past decade rather than steeply declining as one would expect. Today it is responsible for 42 percent of adult deaths, almost four times the rate of 1900, when heart disease is believed to have caused just 8 percent of deaths[6] and was the fourth leading cause of death.[7] In real numbers, far more Americans are dying from heart disease than ever.

Physical activity is undeniably an important factor in modern chronic diseases. With the advent of cars, televisions, and computers, Americans have become sedentary. While today just 20 percent of people have jobs that involve strenuous physical labor, in the mid-20th century it was around 50 percent.[8] Globally, physical inactivity has had as deleterious an effect on health as smoking and obesity, says Dr. I-Min Lee, a professor of epidemiology at the Harvard School of Public Health, who led a recent study on the effects of inactivity.[9] However, research also suggests much of the US shift toward sedentary lifestyles had already occurred by 1960 or 1970.[10] So lack of exercise certainly does not fully explain the latest, worrying trends.

Why in recent decades has an obesity epidemic emerged, and why do heart disease and other chronic diseases continue to widely plague us despite changes that should have dramatically reduced them? The evidence points to diet. But not in the ways people often expect.

How Our Diet Has Really Changed

For the past 50 years we've heard many times diet is a linchpin to modern Western health problems. Author Michael Pollan has aptly described the situation this way: "Scientists operating with the best of intentions, using the best tools at their disposal, have taught us to look at food in a way that has diminished our pleasure in eating it while doing little or nothing to improve our health."[11]

Over and again we've been told a single, simple message: What's wrong with our diet boils down to the butter on our toast, the cream in our coffee, the steak we're having for dinner. Nearly universally, public health agencies, dietitians, and doctors have advised us in the past several decades to cut back on red meat and fats, especially animal fats.[12]

The origins of this advice can be traced back to a 1953 study often called the Seven Countries Study,[13] conducted by Ancel Keys, a University of Minnesota epidemiologist. Keys believed saturated fat was the primary cause of heart disease. The idea made him famous (Keys graced *Time* magazine's cover in 1962) and changed the eating habits of generations to follow. Even without ever having heard of Keys in my youth, I'd heard his idea so often I considered it incontrovertible truth in college, when I decided to adopt a vegetarian diet.

Millions of others also took the message to heart. For the second half of the 20th century, especially after 1970, Americans followed the advice they were hearing everywhere and cut back on red meat and animal fats. We cooked and baked with vegetable oils. (I clearly remember my parents switching from sticks of butter to tubs of corn oil margarine because they believed it to be healthier.) Americans also dutifully ate more fruits, vegetables, and whole grains. With a few notable exceptions, most people I talk to today still consider it gospel that healthy eating means avoiding fats and red meats.

Yet over those decades, America's dietary shift did not improve its collective health. In fact, this was a time period when it got considerably worse. Obesity and hypertension soared, and heart disease and stroke rates persisted. The advice we heard for decades—reducing fats and red meats in our diets would improve our health—was proving wrong in the real world.

Largely because the ineffectiveness of the recommended diet is now so readily apparent, medical and public health researchers and practitioners have been reexamining the standard dietary advice. And a sea change in thinking is under way. A few brave doctors and scientists have begun publicly acknowledging that Americans have been getting bad dietary advice for the past half century.

Momentarily, I'll discuss in some detail what the medical and public health research really shows about linkages between various things we eat and drink and the current health crisis. But first, it's important to dive into some specific data about how our eating patterns have *actually* changed over the past century, especially in recent decades.

Consumption levels of just about every food can be determined by looking at relatively undisputed government data. For over a century, the US Department of Agriculture (USDA) has tracked the amounts of everything we eat and drink.

The first point to note is Americans are simply eating a lot more these days. Between 1970 and 2003, average daily caloric intake increased by a staggering 523 calories per person.[14] Theories abound as to the cause of this increase. Regardless of the reason, this fact alone certainly explains part of the problem.

Equally troubling is the source of those additional 523 calories. Fast food now accounts for a third (34 percent) of the calories Americans consume. Snacking accounts for more than half of the recent calorie increase.[15] At the same time, the amount we take in at dinner has actually declined. In other words, more Doritos, fewer rutabagas. Or perhaps, more accurately, more doughnuts, less steak.

Which leads to an important first question: Are we really eating more red meat and animal fat than ever?

In a word: no. In fact, the reverse is true. We are eating less of both.

During the 20th century American meat consumption fluctuated. Due to wartime rationing, there was a dip in the 1940s. (Sugar was also rationed, which, as will become clear shortly, is noteworthy.) Around 1970 (most likely due to uncharacteristically low prices at the meat counter resulting from overproduction), there was an upward bump of beef eating. But over the course of the century, and especially for the past three decades, the general trend for both red meat and animal fats was downward. Both are notably lower today than they were a century ago.

Specifics are helpful here. (All figures I cite for farming and food production and consumption, except where otherwise indicated, are from electronic versions of original US Department of Agriculture records that I have personally examined.) While Americans ate 71 pounds of beef per person per year in 1905, they ate 60 in 2010. In those same years, veal consumption went from 7 pounds per person to just 0.4 pound. Lamb consumption went down from 6 pounds to just 1. And pork consumption went from 62 pounds to 48. In a similar pattern, egg consumption went from 284 per person to 243. All significantly down. Stated another way, compared with a century ago, we are now eating 11 pounds less beef per year, 6.5 pounds less veal, 5 pounds less lamb, 14 pounds less pork, and 41 fewer eggs.[16] We are not a nation of people who have increased animal fat and red meat consumption.

Now let's look more closely at recent decades. In that time period, red meat consumption went strongly down, especially beef. From 1970 to 2005, beef consumption decreased by 22 percent; pork went down by 3 percent. Overall, red meat consumption decreased by 17 percent.

For foods rich in saturated fats, the trend is even clearer. From 1970 to 2005, butter consumption declined by 15 percent; lard went down by 47 percent; and whole milk consumption plummeted by 73 percent. The only major exception to this pattern was cheese (from pizza and other fast-food consumption), which increased.

The rise in cheese eating, however, was not enough to counteract an overall consistent shift toward less consumption of animal fats. Mary Enig, a PhD food scientist who specialized in fats and authored *Know Your Fats*, calculated that over the course of the 20th century Americans' overall consumption of saturated fats went down by 21 percent.[17]

At this point, we may well wonder how, in the face of these consumption facts, anyone could blame America's health crisis on animal fats and red meat. Indeed, if we believe in the Keys hypothesis, this data is deeply confounding.

But things get still sketchier. Changes in red meat and fats are only part of the story of our shifting eating and drinking patterns. As our consumption of animal fats went down, our vegetable fats went way up. Total per-person cooking and baking fats more than doubled from 1909 to 2009, from 36 pounds to 80 pounds, increasing by 63 percent from just 1970 to 2000. From 1910 to 1970, vegetable fat consumption had already gone up by over 400 percent.[18] For a lot of people, like my parents, this change was due to Americans trying to follow health advice to replace animal fats with vegetable oils. The most common vegetable oils cost less, too, which has certainly been another part of the explanation for the shift. Whatever the reasons, the data clearly show Americans have broadly substituted vegetable oils for animal fats.

Similarly, as physicians and nutritionists warned us to get away from red meat, Americans dutifully migrated away from pork, beef, veal, and lamb, and began eating substantially more chicken, turkey, and fish. In 1909, Americans ate 15 pounds of chicken per year. By 2009, that had gone up to 80 pounds—a fivefold increase. Average turkey consumption went from 2 pounds in 1935 to 9 pounds in 2009—a fourfold increase.[19] Americans also chose to eat more fish. From 1910 to 1999, per-capita consumption rose from 11.2 to 15.3 pounds per person per year (a 37 percent increase).[20]

If red meats are the problem, and poultry and fish are so much better for us, why have all the major diet-related diseases gotten worse as we've eaten more of the "white meats" and cut back our lard, dairy fat, and red meat?

And what about those vegetable fats? Aren't they supposed to be so much healthier than fats from animals? That advice may turn out to have been the worst of all. For one thing, a lot of the vegetable fats Americans used in the 20th century were partially hydrogenated vegetable oils—in other words, the odious manmade trans fat.

"Trans fat is formed during food processing when some hydrogen is added to vegetable oil to increase solidity," explains an article in *Scientific American*. "It is typically added to food to increase its shelf life, improve its texture and maintain its flavor."[21] Long shelf life! Then it can sit in a warehouse or vending machine for months before we eat it! Great.

Crisco kicked things off in 1911, introducing a shelf-stable fat for baking. It was the first time manmade trans fat came into the American diet. Over the 20th century, trans fat became a ubiquitous staple, found in everything from potato chips to Oreos and Girl Scout cookies. I vividly remember TV ads in my childhood featuring Loretta Lynn vouching for Crisco to make "the flakiest piecrusts." (As an added bonus, it needed no refrigeration and could sit in your cupboard for months.)

Ironically, consumer advocacy groups lobbied fast-food restaurants to replace beef tallow and tropical oils with partially hydrogenated oils containing trans fat. McDonald's stopped using beef tallow for its once tasty french fries. It was immediately obvious to many that this change lessened the fries' palatability. In hindsight, it also injured the health of countless diners.

Due to scientific research connecting it to heart disease, Denmark began strictly limiting manmade trans fat in foods in 2003. Other European countries and some Canadian cities and provinces followed suit. When Mayor Bloomberg pushed through a trans fat ban for New York City restaurants in 2006, some of my New York friends said he was a crusading public health zealot who was stepping over the line.

But history has vindicated Bloomberg's actions on trans fats. Earlier that year, the US federal government had begun requiring labeling of foods containing artificial trans fat. And in November 2013, the FDA tolled the death knell for artificial trans fat by announcing it should eventually be entirely removed from foods. In so doing, FDA noted that the US Centers for Disease Control and Prevention had estimated getting rid of trans fat would prevent an additional 7,000 deaths from heart disease annually and up to 20,000 heart attacks every year.[22] "There can be little doubt that the presence of laboratory manufactured trans fats is part of reason for the rise in chronic diseases," stated the FDA announcement. Noting part of its responsibility is to ensure a safe food supply, FDA announced it had made a preliminary determination that trans fat no longer met the standard of "generally recognized as safe." FDA subsequently confirmed this determination. On June 18, 2018, foods containing artificial trans fat (or, as FDA refers to it, partially hydrogenated oils) were classified as adulterated and became illegal.[23]

Unfortunately, the government's action on manmade trans fat, while laudable and necessary, came a century too late. Tens of thousands of people who ate cookies, pies, and french fries containing artificial trans fat made from vegetable fats now suffer, or one day will suffer, cardiovascular disease and heart attacks as a result. Clearly, manmade trans fat partly explains why heart disease rates rose over the 20th century and have stubbornly refused to go down.

It's worth noting the work and life of the leading US researcher on the health effects of trans fats, Dr. Fred Kummerow. He warned of their dangers decades ago, and was one of numerous prominent scientists who have long and vociferously questioned Ancel Keys's hypothesis that saturated fats cause heart disease. Despite the recent acceptance of Kummerow's warnings on trans fats, he continued to consider the public health message about fats totally wrong.[24] Kummerow steadily conducted research on fats since he was a young nutrition PhD scientist at University of Illinois in 1957. Still actively researching until shortly before his death at age 102, he always maintained there is nothing inherently unhealthy about the natural saturated fats found in foods from animals. (It's noteworthy, too, that he was a lifelong red meat eater, and ate eggs fried in butter every morning for breakfast.)

What matters, Kummerow explained, is not whether fat is saturated but whether the fat is oxidized. "Cholesterol has nothing to do with heart disease, except if it's oxidized," Dr. Kummerow said in a 2013 *New York Times* interview. High temperatures used in commercial frying cause polyunsaturated vegetable oils to oxidize, while soybean and corn oils can oxidize inside the body. "If true," the *Times* article notes, "the hypothesis might explain why studies have found that half of all heart disease patients have normal or low levels of LDL [low-density lipids, often referred to as "bad cholesterol"]."[25] In fact, a nationwide analysis done in 2009, done the year following the study cited here by the *Times*, found an even greater disparity—nearly 75 percent of patients hospitalized for a heart attack had cholesterol levels that would indicate they were not at high risk for a cardiovascular event, reinforcing the questionable nature of this purported link.[26] (Interestingly, as I'll discuss later, very-high-temperature food preparation may play a substantial role in turning otherwise healthy foods into foods that, over time, cause health problems.)

The most troubling aspect of the whole manmade trans fat debacle is that it was so unnecessary. Tens of thousands of people have almost certainly suffered and died because they ate it. Yet none of it needed to happen. Trans fat served no noble purpose like making food more nutritious or healthier (or even tastier!); nor was it created to feed starving people. Rather, it was simply a laboratory-produced substance to give food added shelf life. Is it any surprise that such an item would make us sick?

The situation fits perfectly into a general rule suggested decades ago by British physiologist, professor of medicine, and clinical physician John Yudkin (about whom I will say more later). Dr. Yudkin spent his career conducting epidemiological and clinical research on the effects of various foods on the human body. His particular interest was heart disease. Decades of investigation led him to reach the following eminently sensible conclusion: Any food humans and their ancestors have been eating for millions of years is unlikely to cause heart disease or other serious, chronic diseases. He considered meat, fish, fruit, and vegetables to be among the inherently *un*suspicious foods.

The converse, Yudkin urged, is equally true: Any food that humans and our pre-human ancestors did *not* eat during their millions of years of evolution should be considered suspect.[27] Sucrose (sugar) is an example of such a food. Such a simple, sensible idea. Applying Dr. Yudkin's principle, laboratory-generated trans fat should never have been widely introduced into the food chain until long-term studies had proven it safe. Had we taken such an approach, the health and lives of tens of thousands could have been spared.

Yudkin's idea—foods our bodies evolved with are innocent until proven guilty—may have been deemed quaint by some of his peers in the 1960s and '70s. It was the era when space-age foods like Tang and sugar-loaded treats like Twinkies were being invented. Today, however, it is being taken seriously by leading scientists and physicians. Momentarily, I'll discuss Dr. Yudkin's main research findings and recent studies that backs them up.

Doubts and Proofs

First, though, I want to return to the overall question of what proof actually exists to connect red meat and saturated fats to the current epidemic of obesity, hypertension, and chronic diseases. There are several broad categories of studies used to determine what causes chronic diseases.

The food consumption data we've been reviewing thus far would be used in epidemiological studies. It clearly shows that, compared with a century ago, Americans are eating less saturated fat, less red meat, and fewer eggs. Instead, we are eating more vegetable oil, fish, and poultry.

USDA consumption data also shows that over the past three decades we've been eating more whole grains, fruits, and vegetables. Additionally, we've seen that other factors that significantly affect chronic disease rates and their complications (smoking rates, medical care) have changed in ways that should lead to lower rates of disease and fewer deaths. Yet all of these positive changes have failed to counteract the worrisome health trends, at least in a meaningful way. Something is clearly amiss.

On its own, of course, food consumption data cannot be used to prove one food or another is causing America's health problems. Such data helps devise theories; it helps draw up a list of suspects; it's one piece of the puzzle. As a former assistant district attorney, I know in a courtroom you must prove a criminal defendant's guilt beyond a reasonable doubt. A defendant, meanwhile, need not prove his innocence. The accused merely needs to pose unanswered questions—to raise doubt—about his guilt.

As a trained biologist, I am also familiar with the methods used to test a scientific hypothesis. As in law, declaring a hypothesis sound carries a much heavier burden than merely poking holes in one. There is (or should be, anyway) a high bar for condemning a particular food, stating it causes chronic diseases. Taking a cue from Dr. Yudkin, I would add that the burden of proof should be especially high when the food in question has been part of the human diet for millions of years. While proving with certainty the cause of any long-term illness may be nearly impossible, there are various forms of evidence that, in combination, can reasonably be deemed proof.

If red meats and animal fats now stand accused (and there's little doubt they *do*), Ancel Keys is their original accuser. It's easy to see the consumption data raises serious doubts about his hypothesis. Said another way, the consumption data would be insufficient to affirmatively *prove* red meat and fats are good for you, but it is ample to raise serious *doubt* about whether those foods are guilty of causing America's poor health, as has been so commonly alleged.

Taking this analogy a bit further, the Keys "Seven Countries" study could never meet the standard of proof needed to establish the guilt of red meat or saturated fat. Nor, from a scientific view, is it sufficient to prove his hypothesis. Keys merely looked at two conditions existing in the same country (collective saturated fat consumption levels and heart disease rates) and noted a correlation. He then inferred causation. This falls squarely into the category of the sort of epidemiological study that should trigger various forms of additional investigations to test a hypothesis.

This is not to suggest that observational, ecologic research like Keys's study does not have value. It can be invaluable in certain situations—for example, if you're trying to figure out where an infectious disease outbreak is originating. Even without interviewing the afflicted, merely plotting their physical locations could be crucial in coming up with a list of suspect sources for the pathogen. In contrast, when attempting to prove causation for a chronic disease from a particular food item, Keys's type of epidemiological research is extremely weak and should only ever have been a starting point.

Here's why. Keys showed nothing whatsoever about the diets of the individuals in the seven countries. He had no way of knowing whether *any* of the people, let alone *all* of the people, eating higher levels of saturated fats were the same people who were getting sick with cardiovascular disease. Theoretically, it is possible *none* of the people eating large amounts of saturated fat had heart disease. It is equally possible people who ate little saturated fats were those with the disease. All Keys showed was that in seven countries the *total amount* of saturated fat consumed correlated, to an extent, with the *total presence* of heart disease.

Keys's conclusion is further weakened by his own methodology. The same type of data he used for the 7 countries was available for at least 22 countries. He chose only data from the seven countries that most strongly supported his thesis.[28] This is just bad science. When data from all 22 are graphed, the correlation appears weak indeed. For all these reasons, the Keys study, on its own, is poor-quality evidence. Certainly, it is insufficient to convict saturated fat of the crime of causing America's current epidemic of poor health.

The other major piece of evidence taken as proof of the saturated fat hypothesis relates to cholesterol. Clinical trials in the 1970s, following publication of Keys's theory, showed that eating saturated fat raises levels of low-density lipid cholesterol (commonly referred to as LDL cholesterol, or just LDL). This phenomenon is now quite well documented. As I'll describe in more detail momentarily, it is now recognized the research has serious limitations. In the early days of cholesterol research, however, few nuances were understood. Saturated fat's effect on LDL was deemed further proof it caused coronary disease. The matter was treated as a case closed.

Lately, however, the case against saturated fat and red meat is being reopened. Three things seem to have changed. First, it has become increasingly apparent there were problems with the original research, such as those just described. Second, there have been serious flaws underlying subsequent studies that seemed to link beef and saturated fat to health problems. And third, as I'll talk about more in a bit, new research (and some older research) links chronic health problems to other foods and other categories of food.

One serious shortcoming with health studies of diet is what's called the *healthy user bias*. Chris Kresser is an author and integrative medicine practitioner in Berkeley, California, who considers both beef and fat important components of a health-promoting diet. I have met Kresser, and even sought him out last year for his professional counsel on nutrition as I contemplated changing my own diet. Kresser explains the healthy user bias like this: "Since red meat has been vilified for years in the mainstream press, people who eat less of it are also more likely to eat less of other foods that *are* actually unhealthy (i.e. refined sugar, trans fats, processed foods, etc.) and to engage in healthier lifestyle choices (i.e. they are physically active, don't smoke, etc.)." Kresser also points out the types of questionnaires used to track food intake are notoriously unreliable. "Do you remember what you ate for lunch last Tuesday? Neither do I." Combined, such flaws in the research have created a body of unreliable health studies on red meat.[29]

Precisely because these types of flaws are so problematic, there is even serious reconsideration of whether diet-related observational medical studies should be used at all. John Ioannidis, MD, professor of medicine and of health research and policy at the Stanford University School of Medicine, is among those spurring this reassessment. In a major profile piece in 2010, *The Atlantic* called Ioannidis "one of the world's foremost experts on the credibility of medical research," and says he is "one of the most influential scientists alive." His fame comes from his work showing "that much of what biomedical researchers conclude in published studies—conclusions that doctors keep in mind when they prescribe antibiotics or blood-pressure medication, or when they advise us to consume more fiber or less meat, or when they recommend surgery for heart disease or back pain—is misleading, exaggerated, and often flat-out wrong." Perhaps more surprisingly, *The Atlantic* also reported the following: "His work has been widely accepted by the medical community; it has been published in the field's top journals, where it is heavily cited; and he is a big draw at conferences."

One of Dr. Ioannidis's main areas of interest has been the field of nutrition. To the constant barrage of diet-related studies with conflicting results, he suggests a simple, elegant response: "Ignore them all." His reasons for skepticism are multifold. First, in any large database of nutritional and health factors, there will be apparent connections that are in fact merely flukes, not real health effects. Second, even if a study correctly identifies a genuine health connection to some nutrient, a person is unlikely to benefit much from taking more of it. Nutrients act together as a sort of network, Ioannidis notes, and taking more of a particular nutrient is bound to cause effects to the network that these studies cannot detect. Adding a single nutrient "may be as likely to harm you as help you," he believes. Third, because few dietary studies last long enough, there's very little information about longer-term effects for any dietary change. These shorter-term studies measure health "markers," like cholesterol levels. But "meta-experts have shown that changes in these markers often don't correlate as well with long-term health as we have been led to believe."[30]

On top of all of that, Ioannidis sees additional serious problems. In those few studies that last long enough to track death rates, he says, the findings frequently conflict with results of the shorter studies on the same questions. All of these issues are in addition to "ubiquitous measurement errors (for example, people habitually misreport their diets in studies), routine misanalysis (researchers rely on complex software capable of juggling results in ways they don't always understand), and the less common, but serious, problem of outright fraud (which has been revealed, in confidential surveys, to be much more widespread than scientists like to acknowledge)." Finally, even if a study succeeds in correctly identifying a real connection between diet and health, no individual is guaranteed to benefit from changing their diet. "[S]tudies report average results that typically represent a vast range of individual outcomes . . . [S]tudies usually detect only modest effects that merely tend to whittle your chances of succumbing to a particular disease from small to somewhat smaller," *The Atlantic* article notes. Overall, Ioannidis says, "The odds that anything useful will survive from any of these studies are poor."[31]

In a 2018 interview with *Stanford Medicine News*, Dr. Ioannidis summed up his opinion on the low value of diet-related epidemiological health studies. He said:

> *We still largely depend on nonrandomized studies to assess questions of nutrition. These studies are notoriously incapable of giving reliable answers due to confounding factors.* In nutrition, the situation is made even worse because our ability to measure diet is still limited in accuracy, and recall biases, in which study participants remember something incorrectly, can be severe. *In addition, dietary intake of a single nutrient probably has small or even tiny effects on major health outcomes, even if diet as a whole is important. Therefore, any potential finding is largely shaped by the noise from errors and biases of observational studies . . . The biggest problem is that the vast majority of studies are not experimental, randomized designs. Simply by observing what people eat—or even worse, what they recall they ate—and trying to link this to disease outcomes is moreover a waste of effort.* These studies need to be largely abandoned. *We've wasted enough resources and caused enough confusion, and now we need to refocus. Funds, resources and effort should be dispensed into fewer, better-designed, randomized trials.*[32]

Clearly, this entire field of research must be taken with an extra-large grain of salt.

Meanwhile, numerous major studies (as credible as those claiming to find a link, in any case!) have found no link between saturated fats and/or red meat and human diseases. Simultaneously, a growing body of research is documenting that things other than fat and red meat are more likely to be responsible for today's widespread health problems.

Here are some of the studies and re-analyses of the data that have chipped away at the widespread acceptance of the claim red meat and saturated fat cause heart disease.

A 1998 meta-analysis published in the *Journal of Clinical Epidemiology* reviewed dozens of studies including ecological, dynamic population, cross-sectional, cohort, and case-control studies, as well as controlled, randomized trials of the effect of fats in the diet.[33] In particular, it examined the role of saturated and polyunsaturated fatty acids in cardiovascular disease, noting that a diet rich in saturated fats and low in polyunsaturated fats "is said to be an important cause of atherosclerosis and cardiovascular diseases (CVD)." For each study type, the review found highly contradictory study results. Even the type of evidence made famous by Ancel Keys—positive correlations between national intakes of total fat and saturated fat and cardiovascular mortality—were "absent or negative in the larger, more recent studies," the analysis concluded. Overall, the study found a lack of evidence linking saturated fat and coronary disease in any type of study.

This was followed by a 2010 study published in the *American Journal of Clinical Nutrition* and conducted by physicians at the Oakland Research Institute, a biomedical research institute affiliated with University of California–San Francisco. It was a meta-analysis of 21 studies (including 347,747 subjects) to evaluate the association of saturated fat with cardiovascular disease. Noting "a reduction in dietary saturated fat has generally been thought to improve cardiovascular health," the research aimed to "summarize the evidence related to the association of dietary saturated fat with risk of coronary heart disease (CHD), stroke, and cardiovascular disease (CVD; CHD inclusive of stroke) in prospective epidemiologic studies." Using the 21 studies and a random-effects model, it derived composite relative risk estimates for CHD, stroke, and CVD. The researchers found no credible evidence of a link. The study's authors concluded: "Intake of saturated fat was not associated with an increased risk of CHD, stroke, or CVD."[34]

Also in 2010 came a major study by researchers from Harvard School of Public Health.[35] They found that eating processed meat (like bacon and baloney) was associated with a slightly higher risk of heart disease and diabetes. However, they found *no increased risk at all* from eating *unprocessed* red meats, including beef, pork, and lamb. Researchers noted that while dietary guidelines tend to recommend reducing meat consumption, "prior individual studies have shown mixed results for relationships between meat consumption and cardiovascular diseases and diabetes." The Harvard team concluded the reason for the mixed results was that prior studies had failed to separately consider the health effects of processed versus unprocessed red meats. For its analysis, the team began by systematically reviewing nearly 1,600 studies. Twenty relevant studies that distinguished processed from unprocessed meats were identified. Subjects included a total of 1,218,380 individuals from 10 countries on four continents. Neither the fat nor the cholesterol content of the processed meats varied from the unprocessed, the researchers pointed out. Rather, the key difference was the sodium and nitrate content. This led the study authors to conclude: "[D]ifferences in salt and preservatives, rather than fats, might explain the higher risk of heart disease and diabetes seen with processed meats, but not with unprocessed red meats."

Yet another nail in the coffin of the alleged link between saturated fat and heart disease came in early 2014. In what *The New York Times* deemed "a large and exhaustive new analysis by a team of international scientists," a report published in the journal *Annals of Internal Medicine* evaluated the evidence for the link between saturated fat and heart disease. The researchers based their conclusions on nearly 80 studies involving more than half a million people. They looked at what people reportedly ate, as well as more objective measures such as the composition of fatty acids in their bloodstreams and in their fat tissue. Additionally, the scientists reviewed evidence from 27 randomized controlled trials (often referred to as "the gold standard" for health research) that assessed whether taking polyunsaturated fat supplements, such as fish oil, promoted heart health. The extensive review of the medical research found "no evidence that eating saturated fat increased heart attacks and other cardiac events."[36]

Additionally, the review found higher consumption of monounsaturated fats (like olive oil) and polyunsaturated fats (like corn oil) did *not* lessen heart disease risk. "The new findings are part of a growing body of research that has challenged the accepted wisdom that saturated fat is inherently bad for you," the *Times* article continued. Summarizing the findings of this research, Dr. Rajiv Chowdhury, a cardiovascular epidemiologist at Cambridge University and the lead author of the study, said: "It's not saturated fat that we should worry about."

On top of the types of problems highlighted by Dr. Ioannidis and others, the entire body of studies condemning saturated fat was effectively called into question by the book *The Big Fat Surprise* by former National Public Radio journalist Nina Teicholz. The book was named one of *The Economist*'s Books of the Year and *The Wall Street Journal*'s

Top Ten Best Nonfiction Books of 2014. *Kirkus Reviews* called it "solid, well-reported science."

Teicholz's exhaustive review of the science of dietary fat concludes: "What we believed to be true—our conventional wisdom—is really nothing more than sixty years of misconceived nutrition research." She goes on to say,

> *[A] higher fat diet is almost assuredly healthier in every way than one low in fat and high in carbohydrates. The most rigorous science now supports this conclusion . . . Over the past decade, a stack of top-rate scientific studies attesting to the importance of dietary fat has grown to the point where the accumulated body of evidence in nearly undeniable.*[37]

After the release of her book, I met with Nina in New York City and liked her right away. She impressed me with her intelligence, frankness, and credibility. We shared notes on writing, parenting, and speaking out against mainstream dietary advice. An interesting footnote on my connections with her: A year or so later, I was asked to assemble a panel to discuss the shift away from animal fats for a major US conference hosted by the U.K. organization Sustainable Food Trust (SFT). I immediately contacted Nina about being part of the panel and was delighted she agreed. But a bit later I learned from SFT they had been contacted by a prominent American vegetarian food activist who was upset she was included. Anyone so strongly against mainstream thought on fats should not be part of this conference, this person urged. To their credit, SFT never hesitated and was very pleased to have her there. But it speaks volumes about the prevailing dietary groupthink. Anyone who challenges prevailing norms is aggressively shunned.

From my perspective with a legal and scientific background, Nina Teicholz's work and many recent studies exonerate beef and saturated fat from the accusation they cause heart disease. When considered together with the serious flaws inherent in dietary health studies, the case against red meat and saturated fat falls flat. Some of this research also explains why past studies have reached such faulty conclusions. That research suggests that additives frequently used *with* meat—not meat itself—are to blame for health problems.

Another factor likely confounding study results is how meat is cooked. Famed food scientist Harold McGee eats red meat. But he cautions: "We should . . . prepare meat with care." Certain ways of cooking it create compounds that may pose health risks. It is well known that nitrosamines can be generated by very high heating of nitrate-containing meats (such as some sausages). Additionally, McGee notes, scientists have identified two other families of potentially carcinogenic chemicals that can be created by popular cooking methods. One is heterocyclic amines (HCAs), which can be formed at high temperatures by the reaction of creatine and creatinine (compounds present in small amounts in meat) with amino acids. Second, polycyclic aromatic hydrocarbons (PAHs) are created when organic material is heated to the point that it begins to burn. Cooking over a smoky wood fire therefore deposits PAHs from the wood onto the meat (as it would, it should be noted, on any food cooked over an open flame).[38] This line of research suggests it's probably wise to regard using your grill as a treat, rather than a frequent cooking tool. Meat has been a focus of the research merely because meat is what people most often cook on their grills. But here again, meat itself is not the problem. Beef that is unprocessed, not burned, and not prepared over an open flame raises none of these health concerns.

Now let's return to how saturated fat affects LDL cholesterol levels. As mentioned earlier, clinical trials have shown that saturated fat raises LDLs. In the 1970s, this was considered highly damning for saturated fats. Now we know better. The current understanding of cholesterol is far more sophisticated.

According to popular myth, high cholesterol levels cause heart attacks. Yet research shows that 75 percent of heart attack patients have normal or low LDL cholesterol levels.[39] In other words, only one-quarter of people who have heart attacks actually have high levels of LDL. This fact alone shows there must be much more to the heart disease story than elevated LDL levels. It is now widely recognized a connection between saturated fats and LDL levels is not proof of a link to *coronary disease*. Said another way, having an elevated LDL level is not an illness. Rather, LDL levels are simply biomarkers—things medical professionals use to help predict whether or not someone might develop a disease. The LDL number is just one piece of information needed for a patient's cholesterol numbers to serve as a useful biomarker. For one thing, levels of HDL (high-density lipids), the other major type of cholesterol, and the ratio of LDL to HDL are now considered at least as important.

Clinical investigations into the effects of diet on LDL and HDL levels have revealed a complicated picture. Researchers in the Netherlands worked to clarify it by reviewing results from 60 human dietary clinical trials.[40] Noting "the ratio of total to HDL cholesterol is a more specific marker of CAD [coronary artery disease] than is LDL cholesterol," the study evaluated the effects of individual types of fats on cholesterol ratios. The study did find that saturated fats increased LDLs. But, "not by much," Dr. Dariush Mozaffarian, cardiac epidemiologist and faculty member at Tufts University School of Medicine, said of this research. Equally important, the study found that if *carbohydrates* replace dietary saturated fats, the cholesterol ratio was unchanged. The results also showed that replacing fats with carbohydrates actually *increased* fasting triglyceride concentrations. Overall, the study concluded: "The effects of dietary fats on total:HDL cholesterol [the ratio of total cholesterol to HDL cholesterol] may differ markedly from their effects on LDL." By looking at LDLs, previous research may have entirely missed the boat. The total:HDL could have been telling a completely different story.

The latest lines of cholesterol research also show that the specific type of LDL matters. A rise in the level of the larger, fluffier type of LDL cholesterol is now understood to be fine. On the other hand, a rise in the smaller, denser LDL type is worrisome. Recent studies have made the crucial determination that the LDL rise associated with saturated fats is the lighter, fluffier variety. So even that is of much less concern than once believed.[41]

The connection between our diets and cholesterol levels is an evolving story. As the Netherlands research suggests, there is growing doubt about the role of fats. The focus is shifting to carbohydrates, which now appear to matter more than fat. A 2014 *New York Times* article, based on an interview with Cambridge University cardiac epidemiologist Dr. Rajiv Chowdhury, notes that "the relationship between saturated fat and LDL is complex." In addition to raising LDL cholesterol, saturated fat also increases HDL, the so-called good cholesterol. And, Chowdhury notes, "the LDL that [saturated fat] raises is a subtype of big, fluffy particles that are generally benign." The *Times* article summarizes the current science this way: "The smaller, more artery-clogging particles are increased not by saturated fat, but by sugary foods and an excess of carbohydrates, according to Dr. Chowdhury. 'It's the high carbohydrate or sugary diet that should be the focus of dietary guidelines,' he said. 'If anything is driving your low-density lipoproteins in a more adverse way, it's carbohydrates.'"[42]

On top of that, research over the past decade shows that having a high HDL number is actually desirable. A high HDL is now considered protective against heart disease. For women, in particular, an elevated HDL number is more predictive that a person will *not* get heart disease than a high LDL number is that she *will*. And fats, including saturated fats, actually raise HDL. In fact, according to cardiologist Dr. Mozaffarian: "The best way to raise HDLs is with saturated fat."[43]

To understand how the body responds to fats compared to carbohydrates, the work of Dr. Jeff Volek is instructive. A registered dietitian and professor in the Department of Human Sciences at Ohio State University, Volek's work has focused for two decades on understanding the human body's response to carbohydrate consumption. "People believe 'you are what you eat,' but in reality, you are what you *save* from what you eat," he says. "[T]he primary regulator of what you save in terms of fat is the carbohydrate in your diet," Volek continues. "There is widespread misunderstanding about saturated fat. In population studies, there's clearly no association of dietary saturated fat and heart disease, yet dietary guidelines continue to advocate restriction of saturated fat. That's not scientific and not smart." Volek's research shows that doubling saturated fat in the diet does not increase saturated fat in the blood. Carbohydrate consumption, not fat consumption, is the key factor in determining saturated fat in blood, his research has revealed.[44]

Like nearly everyone in my generation, I grew up hearing the Keys-type saturated fat hypothesis as gospel. So the idea that fat could benefit cholesterol levels initially struck me as both improbable and bizarre. And it surprised me greatly to learn highly credible physicians and scientists were questioning Keys's hypothesis all along. In fact, though, in the 1950s and '60s, as heart disease rates were rising in the United States, Britain, and other Westernized countries, various theories were developed to explain the rise. Each theory had some epidemiological evidence to back it up.

The saturated fat theory was not necessarily the one with the most science behind it. But for various reasons (including, apparently, Keys's forceful personality), it won the day, eclipsing other competing theories.

One of the leading skeptics of the saturated fat idea was the British physician and physiologist mentioned earlier, John Yudkin. Dr. Yudkin was the highly respected founder of the first university nutrition department in Europe. Like scientists in other parts of the world, he and his colleagues had examined available public health data in search of the cause of rising heart disease rates. But unlike Keys, they found the evidence blaming saturated fat unconvincing.

The More Plausible Explanation: Sugar

Starting in the 1950s, Dr. Yudkin had examined enough epidemiological evidence to feel he knew why heart disease rates were rising in Western Europe and the United States. He believed the rising rates were due to *sugar*, not fat. Unlike Keys, who plotted just 7 countries on a graph, Yudkin looked at the same type of data from 22 countries (including the 7 Keys used). His broader analysis found a much stronger relationship between heart disease rates and high sugar consumption. Recognizing that this research was only a first step, he and his colleagues at the University of London's Queen Elizabeth College soon launched a series of experiments to further investigate.

For more than a decade that followed, Yudkin and his research team conducted scores of experiments to determine how sugar behaves in the body. Feeding sugar (often alongside experiments feeding starches or fats) to humans, rodents, and other animals, they closely tracked its effects.[45]

Here's a sampling of some of the team's findings. In rats, their trials consistently found that sugar consumption increased blood pressure, increased blood level triglyceride, and decreased the body's efficiency in dealing with elevated blood glucose. Sugar was also found to make blood platelets "stickier" (more adhesive to one another and to arterial walls), a precursor to atherosclerosis. Other rat experiments, testing levels of two enzymes present in the livers of rats and humans (pyruvate kinase and fatty acid synthetase) greatly increased (fivefold and twofold, respectively), indicating an increased production of fat in the liver. In various rodents, sugar feeding enlarged the liver by 25 to 100 percent. Several experiments showed increased insulin resistance from sugar. In several species (baboons, chickens, pigs, and rabbits), sugar increased cholesterol levels.[46]

Feeding experiments on human subjects (both in Yudkin's lab and elsewhere) subsequently showed similar effects in humans. Sugar in human diets resulted in elevated cholesterol and triglyceride levels. The trials also found a notable rise in human insulin levels—40 percent after just two weeks on a high-sugar diet. Human adrenal hormones also rose significantly, by 300 to 400 percent.

As for sugar's mechanism, Dr. Yudkin acknowledged uncertainty. In trying to understand how sugar could be involved with so many diseases, he felt two overriding points were key. "One is that sugar produces an enlargement of the liver and kidneys of our experimental animals, not only by making all the cells swell up a little, but by actually increasing the number of cells in these organs"—conditions referred to as hypertrophy and hyperplasia.[47]

The second effect he considered particularly significant was that in some people sugar produced an increase in the levels of insulin, estrogen, and adrenal cortical hormone. Systemic effects like these make it plausible for sugar to be implicated in a large number of conditions. And such effects on the hormones and vital organs, he believed, "should persuade any reasonable person that sugar is not just an ordinary kind of food."[48]

For atherosclerosis, the disease in which plaque builds up on the walls of arteries, blood vessels that carry oxygen-rich blood to the heart and other parts of the body, Yudkin offered a "working hypothesis." It was grounded on the belief that the underlying cause is a high level of insulin. Many people known to have hardening arteries also have high blood levels of insulin, he noted. Furthermore, coronary atherosclerosis is accompanied by elevated cholesterol and triglyceride levels, and a range of other disturbances in biochemistry and in platelet behavior. "Only a disturbance of hormone levels is likely to afford an explanation for such a wide variety of changes," he concluded. Various pieces of evidence pointed to insulin as the hormone most likely to be at fault.

Yudkin's investigations were careful, methodical, and empirical. Over the years he published some 300 articles in scholarly journals, including

Britain's leading medical journal, *The Lancet*, reporting the results of his research. Having reviewed all extant medical studies on sugar, he remained skeptical about some claims of its harmful effects (including American research connecting sugar to childhood hyperactivity and learning disorders). And he was convinced that sugar is not the sole cause of heart disease, which he considered a complex illness triggered by multiple factors.

Ultimately, though, Yudkin found the results from his extensive clinical trials strongly supportive of the epidemiological evidence. Collectively, it was compelling proof that sugar was the primary cause of the modern era's global rise in heart disease.

Sugar affects people differently, he concluded. For nearly all people, sugar fosters tooth decay and gum disease. For many, it contributes to dyspepsia (indigestion), as sugar can produce or exacerbate an inflamed mucous membrane of the esophagus or stomach. And for some people—not the majority, but a large portion (25 to 30 percent, Yudkin estimated)—whom he deemed sucrose-sensitive, sugar triggers a broad array of health problems. Among them is heart disease. As with smoking—some lifelong smokers get emphysema, some get lung cancer, and some get neither—only some people who eat sugar at levels typical of the modern diet will become afflicted by heart disease.

Throughout the years of his experiments, Yudkin was a prolific presenter of his findings at nutrition conferences. He published numerous articles about the trials. When he completed his clinical investigations into sugar, Yudkin dedicated a year to summarizing his research and conclusions in the book *Pure, White, and Deadly* (first published in 1972, then updated in 1986). In it, he pointedly questioned the purported links among heart disease, meat, and saturated fat. Foods humans have been eating for eons.

"Our view, then," he wrote, "is that the underlying cause of coronary disease is a disturbance of hormonal balance. Apart from increased insulin and adrenal hormone, for example, many patients show an increase in estrogen." The agent causing the disturbance was sugar.

Yudkin, as the founder and chair of a leading university nutrition department, was prominent in Britain and internationally. The clear and strong condemnations of sugar he made in *Pure, White, and Deadly* were particularly dangerous to a burgeoning processed-food industry. As health officials began warning of the dangers of fat, the industry was already devising "low-fat" food formulations. They were loaded with sugar to replace flavor removed with fat.

Accordingly, the processed-food industry, joined by rival physician Ancel Keys, mounted a campaign to discredit Yudkin and his work. "Keys loathed Yudkin and, even before *Pure, White, and Deadly* appeared, he published an article describing Yudkin's evidence as 'flimsy indeed,'" according to a biographical newspaper article.[49] Despite his decades of rigorous research, sugar industry trade groups issued press releases dismissing Yudkin's statements as "emotional assertions" and describing his book as "science fiction." When Yudkin sued in court, they were forced—nearly five years later—to print a retraction. Of course, by that time the damage to his reputation had already been accomplished.

For the rest of his life, Yudkin suffered personally and professionally for his scientific publications and statements about sugar. The processed-food industry and sugar industry evidently regarded him as their foe. He found himself "uninvited" to international conferences, forced to cancel conferences he had organized, and blackballed from publishing his findings.[50] His own university even reneged on a promise to allow the professor to use its research facilities in his retirement. "Only after a letter from Yudkin's solicitor was he offered a small room in a separate building."

His despair over the situation is palpable in the words of his updated (1986) version of *Pure, White, and Deadly*:

> *Can you wonder that one sometimes becomes quite despondent about whether it is worthwhile trying to do scientific research in matters of health? The results may be of great importance in helping people avoid disease, but you then find that they are being misled by propaganda designed to promote commercial interests in a way that you thought only existed in bad B films.*[51]

Despite industry efforts to discredit him and repress his findings, a number of other scientists heard and read about Yudkin's work. Some took up their own investigations into sugar. Dr. Richard Ahrens, a nutrition professor at University of Maryland who had spent time in Dr. Yudkin's lab, performed US experiments demonstrating that sugar causes an increase in blood pressure in both rats and humans. Dr. Ahrens's experiments and epidemiologic research led him to conclude that sugar, not fat, was the primary trigger of heart problems in the developed world. Coronary heart disease has increased, he noted, "on a world-wide scale in rough proportion to the increase of sucrose consumption but not in proportion with saturated fat intake."[52]

Similarly, Dr. Sheldon Reiser, at the US Department of Agriculture Nutrition Lab in Beltsville, Maryland, conducted human feeding trials with sugar. The research found about 25 percent of subjects to be sugar-sensitive. The elevated insulin levels with sugar consumption were commensurate with the typical American diet. Dr. Reiser's experiments also confirmed that adding sugar to human diets resulted in increased blood levels of triglyceride, cholesterol, and glucose. Even a sugar-industry-funded study (conducted at Wake Forest University in North Carolina) found that pigs fed sugar had elevated cholesterol levels and the genesis of atherosclerosis.[53]

One might expect the collective weight of such scientific findings to have catalyzed a major push within the medical research community to further investigate sugar's effects on the human body, especially as it relates to coronary disease. Perhaps meat and saturated fats had been wrongly fingered? Instead, Yudkin's line of research nearly dried up and withered away. His public shunning seems to have had a serious chilling effect on other medical researchers. By the end of the 1970s, a recent biographical article states, "he had been so discredited that few scientists dared publish anything negative about sugar for fear of being similarly attacked."[54]

Part of the problem for Dr. Yudkin was that he was so far ahead of the curve. Although his experiments consistently showed detrimental effects from sugar, neither he nor anyone else fully understood the mechanisms. "Three or four of the hormones that would explain his theories had not [yet] been discovered," notes David Gillespie, author of the book *Sweet Poison*.

Epidemiological and clinical research done over the past decade has more than fully restored Yudkin's credibility and breathed new life into his line of inquiry. It has, Dr. Robert Lustig has said, revealed his work as "prophetic." The recent studies go beyond just establishing a link to obesity as a precursor to other diseases. They directly connect sugar consumption to heart disease, diabetes, and other serious afflictions.

Suggesting sugar makes you fat, which in turn makes you unhealthy (including health problems like diabetes, hypertension, and heart disease), has never been terribly controversial. In fact, other than by Yudkin and a few others, that was the standard assessment of sugar until fairly recently. But the idea that sugar has to make people fat to trigger adverse health events has changed. Beyond merely being an empty, excessive calorie, it is now widely recognized that sugar—*whether or not it makes you fat*—is bad for your health.

Here are a few examples. The Nurses' Health Study, which followed nearly 90,000 women over two decades, found that women who drank more than two servings of sweetened drinks a day had a 40 percent higher risk of heart attacks or death from heart disease than women who rarely consumed sweet drinks.[55] Similarly, a study following 40,000 men for two decades found a 20 percent higher risk of having a heart attack or dying from a heart attack for those who averaged one can of a sugary beverage per day compared with men who rarely or never drank them.[56] Another study found people who regularly consume one or more sugary drinks a day to have a 26 percent greater risk of developing type 2 diabetes.[57] And a 22-year-long study of 80,000 women found those who consumed a can of sugary drink a day had a 75 percent higher risk of gout.[58]

The new thinking on sugar was exemplified in the February 3, 2014, issue of the *Journal of American Medical Association Internal Medicine*. Along with a major new study linking sugar to death from heart disease, the journal contained a noteworthy commentary by Laura Schmidt, PhD, of the University of California–San Francisco School of Medicine. In "New Unsweetened Truths About Sugar," Dr. Schmidt describes a sea change in perspectives now occurring with respect to sugar. She writes:

> *We are in the midst of a paradigm shift in research on the health effects of sugar, one fueled by extremely high rates of added sugar overconsumption in the American public. By "added sugar overconsumption," we refer to a total daily consumption of sugars added to products during manufacturing (i.e., not naturally occurring sugars, as in fresh fruit) in excess of dietary limits recommended by expert panels. Past concerns revolved around obesity and dental caries [tooth decay] as the main health hazards. Overconsumption of added sugars*

has long been associated with an increased risk of cardiovascular disease (CVD). However, under the old paradigm, it was assumed to be a marker for unhealthy diet or obesity. The new paradigm views sugar overconsumption as an independent risk factor in CVD as well as many other chronic diseases, including diabetes mellitus, liver cirrhosis, and dementia—all linked to metabolic perturbations involving dyslipidemia, hypertension, and insulin resistance. The new paradigm hypothesizes that sugar has adverse health effects above any purported role as "empty calories" promoting obesity. Too much sugar does not just make us fat; it can also make us sick.[59]

The new study accompanying Dr. Schmidt's commentary was carried out by researchers from Harvard and Emory universities and the Centers for Disease Control and Prevention. Called "Added Sugar Intake and Cardiovascular Diseases Mortality Among US Adults," it connects sugar to heart disease deaths. Noting that epidemiologic studies have long shown an association of sugar with cardiovascular disease, the researchers sought to address the relative dearth of prospective studies examining sugar's association with heart disease mortality. They tracked 163,039 "person-years," which included 831 heart disease deaths, during a follow-up of about 15 years.[60] Adjusting for numerous variables, including age, sex, and socioeconomics, the researchers conclude: "The risk of CVD [cardiovascular disease] mortality increased exponentially with increasingly usual percentage calories from added sugar."

More specifically, the study found that the risk of CVD mortality becomes elevated once added sugar makes up more than 15 percent of daily calories. A single can of soda a day increases the risk of cardiovascular diseases by 30 percent.[61] Noting that "[m]ost US adults consume more added sugar than is recommended for a healthy diet," the study concludes: "We observed a significant relationship between added sugar consumption and increased risk for CVD mortality."[62]

More and more, this is even becoming the message of physicians communicating to mainstream audiences, like Dr. Mark Hyman, director of Cleveland Clinic's Center for Functional Medicine, who writes a health column for *The Huffington Post*. "Here's the take-home message," Hyman advises. "Fat doesn't make you fat. Sugar makes you fat." He recommends people eat good-quality, "real, whole" foods (including grass-fed beef), and not worry about fat intake.[63]

Somehow, it's just harder to wrap one's brain around sugar being the true culprit for chronic diseases. For one thing, sugar is so ubiquitous that we tend to assume it must be benign. Fat causing fatness and heart disease also seems so logical. We can easily visualize fat globules floating around in our blood and clogging up our arteries. But sugar? How does that work? The whole idea is much less intuitive.

Despite recent advances in our understanding, the precise mechanism by which sugar works inside the body remains somewhat uncertain. But a great deal of progress has been made since Yudkin's day. Researchers have now established how the body metabolizes the two component carbohydrates of table sugar (sucrose), which are fructose and glucose. It is now known all cells metabolize glucose, while fructose is primarily metabolized by the liver. "This means consuming excessive fructose puts extra strain on the liver, which then converts fructose to fat." A fatty liver, in turn, can cause metabolic syndrome.

Metabolic syndrome is a relatively new term used to describe a group of risk factors for cardiovascular disease that commonly cluster together. A person is generally considered to have metabolic syndrome when she has three or more of the following five criteria: abdominal obesity, raised triglycerides, reduced HDL, elevated blood pressure, and raised plasma glucose. (Several of these conditions, of course, are the very ones Yudkin's experiments connected to sugar decades ago.) Most people with metabolic syndrome will also be insulin-resistant (which Yudkin also showed results from sugar consumption); they may or may not have type 2 diabetes (something that various researchers other than Yudkin had contemporaneously connected to sugar).[64] In addition to cardiovascular disease and type 2 diabetes, individuals with metabolic syndrome are susceptible to other conditions, including polycystic ovary syndrome, fatty liver, cholesterol gallstones, asthma, sleep disturbances, and some forms of cancer.[65]

In a 2014 essay titled "Sugar Is Now Enemy Number One in the Western Diet,"[66] London's Croydon University Hospital cardiologist Aseem Malhotra seemed to inherit Dr. Yudkin's mantle. He is leading the charge to focus the medical and public health communities away from worrying about meat and fats and toward reducing sugar consumption.

His article equates the devastating health effects of excessive sugar with smoking, as it compares the food industry's methods of bullying academic researchers and manipulating the science to the tactics infamously employed by the tobacco industry. Noting that poor diet is actually responsible for *more* disease than smoking, alcohol, and physical inactivity combined, he argues that the bulk of scientific evidence now supports focusing on a particular dietary change: "The evidence suggesting that added sugar should be the target is now overwhelming."

On Medical News Today, Malhotra spoke more specifically about the science behind targeting sugar.[67] Vilification of saturated fat and cholesterol has been misguided, he says. "It is time to bust the myth of the role of saturated fat in heart disease and wind back the harms of dietary advice that has contributed to obesity." Red meat and dairy products, he points out, contain nutrients essential to human health. Additionally, "[r]ecent prospective cohort studies have not supported any significant association between saturated fat intake and cardiovascular risk." To the contrary, Malhotra says, "saturated fat has been found to be protective." Three-quarters of people hospitalized for heart attacks have normal cholesterol levels, he also confirms.

On the other hand, two-thirds of hospitalized heart attack sufferers have metabolic syndrome.[68] The real issue, Malhotra says, is a related phenomenon, a triad of lipid abnormalities called atherogenic dyslipidemia[69] (or atherogenic lipoprotein phenotype), which is characterized by increased blood concentrations of small, dense-LDL particles, decreased HDL particles, and increased triglycerides. "Atherogenic dyslipidemia is characteristically seen in patients with obesity, the metabolic syndrome, insulin resistance, and type 2 diabetes mellitus."[70] Research done by the Cardiovascular Research Center and Center for Human Genetic Research at Massachusetts General Hospital in 2010 showed that "[d]ietary saturated fat content has little influence on the components of the atherogenic lipoprotein phenotype."[71]

Sugar, by contrast, is directly connected to atherogenic dyslipidemia and metabolic syndrome. Regardless of a person's weight, Malhotra says, sugar consumption is an independent risk factor for metabolic syndrome. Although the terms for these medical conditions were coined well after Dr. Yudkin wrote *Pure, White, and Deadly*, their clusters of characteristics bear a striking resemblance to those he described decades ago in his findings on the effects of dietary sugar.

In fact, these days Yudkin's ideas have gained such broad acceptance that in March 2014 the World Health Organization lowered its "target" recommendations for added sugars from 10 percent of dietary calories to 5 percent,[72] which is less than the sugar contained in a single can of soda. WHO noted that its recommendation was based on a wealth of studies now connecting sugar to obesity and chronic health problems.[73] Despite moving in this direction, WHO's then director-general, Margaret Chan, commented when the change was announced on the power of the sugar industry. It seems the sector is not taking such moves lying down, and it will likely continue to fund counter-studies and campaigns to discredit scientists doing sugar-damning research.

Of course, the sugar, beverage, and processed-food industries are all well aware Americans now get about 15 percent of their calories from added sweeteners (those that do not occur naturally in foods),[74] and they are motivated to keep it that way. WHO guidelines propose a huge dietary shift. Dr. Robert Lustig believes it's a change that must occur as the cornerstone in addressing the United States' chronic diseases. Lustig has become something of a folk hero for speaking in clear, unvarnished language about the health dangers of sugar despite the aggressive way such scientists have been attacked by industry in the past. Lustig believes, in addition to obesity and dental problems, sugar is an important factor in heart disease, cancer, Alzheimer's, and diabetes. He also points to science showing sugar is addictive, like tobacco and cocaine.

Dr. Lustig is among the scientists and physicians who now consider metabolic syndrome the key to the major chronic diseases. "Being thin is not a safeguard against metabolic disease or early death," Lustig writes in his book *Fat Chance*. "*The obesity is not the cause of chronic metabolic disease, otherwise known as metabolic syndrome. And it's metabolic syndrome that will kill you.*"[75] Like Yudkin, Malhotra, Schmidt, and others, Lustig does not blame red meat or fat for today's crisis. Sugar, he believes, is the biggest dietary contributor to metabolic syndrome.[76]

And, Lustig argues, there is one simple reason this science has been obscured for so long: the undue influence of the sugar and processed-food industries. Somewhat ironically, his suggestion is only reinforced when you look at Lustig's Wikipedia entry. Nearly two-thirds of the studies cited there to repudiate his views were funded by Coca-Cola.[77]

Having now seen the strong links from sugar to heart disease and other health problems once blamed on red meat and fat, the claim that meat and fat cause today's chronic health problems seems even more implausible. But what does the consumption data show about sugar? Let's now return to the federal government's dietary data and take a look.

The upward trend for sugar consumption is clear and striking. Between 1900 and 2000, consumption of added sweeteners rose steadily—60 percent over the century and 20 percent between 1970 and 2005. By some estimates, the average American now eats 130 pounds of sugar per year, more than 0.3 pound a day, every day.[78] That figure tends to shock people (and, in my experience, people immediately assert their own consumption is much less).

Taking a longer view, though, it's even easier to see the extreme aberrance of today's sugar ingestion. Humans, Dr. Yudkin points out, ate almost no sweetened foods or beverages until quite recently. As he puts it, for 99.9 percent of human history, we ate and drank almost nothing sweet. Sweetness was limited to the fleeting seasons of ripe fruits and the rare treat of wild honey, which is always well guarded by swarms of angry bees.

Sugar as an isolate, an additive, was invented just 2,500 years ago, when a method to extract the sweetness from sugarcane (which, to this day, is no easy task) was devised in India. Still, it remained a rare luxury item for thousands of years.[79] A teaspoon of sugar cost a citizen of Middle Ages Europe as much as a teaspoon of caviar costs us today. For ordinary people, adding a cup of sugar to anything would have been unimaginable. A study on the metabolic effects of sugar in the *American Journal of Clinical Nutrition* notes, "Before the introduction of sugar, the primary sweetener had been honey, but because it was relatively rare and not mass produced, the majority of people (especially the poorer classes) *had no sweeteners at all in their normal diet* even as recently as the early 19th century."[80]

By 1800, per-capita annual American sugar consumption is estimated to have risen to around 6 to 18 pounds. This was much more than in ancient times, but, even at the higher end, only around 15 percent of today's level. Total global sugar production at that time amounted to just 0.25 million ton.[81] By the early 20th century, sugar was quite common. But it was still expensive compared with current prices. In 1920, a 5-pound bag of sugar cost the equivalent of $7.61 (more than double what it costs today).[82]

A graph of US sugar consumption since 1800 is pretty much a line heading steadily upward at about 45 degrees. The World War II years were the exception. Sugar was rationed by the US government from 1943 to 1946,[83] and graphs of US sugar supply dip during those years.[84] Beef, cheese, and other meats were also rationed. Epidemiological studies have sometimes pointed to a subsequent dip in heart disease as a result of rationing those animal-based foods. But isn't it just as likely—in fact, much *more* likely given what we now know—that sugar rationing is the real explanation?

These days, the Food and Agriculture Organization says that by weight sugarcane is the world's largest crop. In 2012, it was cultivated on 31 million hectares (26 million in cane, 5 million in beet), producing 173 million tons of sugar.[85] Per-person consumption in the United States and many other countries is estimated to top 100 pounds per year. In other words, the amount of sugar we eat has exceeded the levels our bodies evolved with for more than two centuries, with a *ninefold* increase over the past 200 years! It is not hyperbole to label current levels, as the above-referenced journal study did, "an epidemic of sugar consumption."[86]

In today's diet, a lot of the sweetness comes from soda and juicy drinks. From 1977 to 2001, US sweetened beverage consumption more than doubled, accounting for 278 additional calories in the daily diet.[87] The average American now drinks over 50 gallons of soda a year.[88] The USC Childhood Obesity Research Center at Keck School of Medicine calls the jump in soda consumption "the prime driver behind the obesity epidemic."[89] Research specifically links sweet beverages to heart disease and adverse changes in lipids, inflammatory factors, and leptin (the hormone that controls weight gain).[90]

Some research suggests the shift to high-fructose corn syrup (HFCS) as a sweetener in beverages and other processed foods has further exacerbated our current health crisis.[91] (Beet and cane sugar consumption have actually slightly declined in the past three decades while consumption of HFCS has spiked a staggering 10,000 percent.) Sucrose, remember, is half fructose, and fructose is now considered the cause of most of sugar's negative health consequences.

Researchers at the University of Southern California recently determined the high-fructose corn syrup in many sodas is as much as 65 percent fructose, nearly 20 percent higher than commonly assumed. "The elevated fructose levels in the sodas most Americans drink are of particular concern because of the negative effects fructose has on the body," explained study author Michael Goran, PhD, professor in the departments of preventive medicine, physiology and biophysics, and pediatrics at the Keck School. "Unlike glucose (the smaller component of HFCS), over-consumption of fructose is directly responsible for a broad spectrum of negative health effects." The body processes fructose differently from glucose, Goran points out. Consuming large amounts of fructose (other than when naturally occurring in whole fruits) greatly exacerbates the risk for those diseases by also causing fatty liver disease, insulin resistance, increased triglyceride levels, and an acute rise in blood pressure—in other words, by bringing about metabolic syndrome.[92]

As I've reviewed the data on rising sugar intake, and the emerging science about the toll it takes on our bodies, my thoughts have often turned to my own father. When he was 42, his parents, within a few months of each other, both died of heart failure. They were in their early 70s. Shaken, my father was determined to have a different fate. Convinced their diets were largely to blame, he immediately swore off sugar, which my father considered a major reason for his mother's obesity and diabetes. A man of extraordinary self-discipline, my father refused sweets for the next 40 years. With the exception of a small slice of home-baked cake on his birthday and a rare piece of homemade fruit pie, the only sweet things he consumed were whole fruits and a small glass of real orange juice with his breakfast.

At some point in high school, I remember gently needling my father about his diet, which seemed to me a bit overboard. Candy was my vice and I always made sure to have a little stash in my backpack (the junkier, the better—Everlasting Gobstoppers and Lemonheads were favorites). Yet I considered my sweet tooth defensible because I was thin and physically active. Taking the conventional view at the time, I told my father: "The only thing wrong with sugar is it doesn't have any nutrients." I remember using a phrase I'd picked up from Jane Brody's *New York Times* health column, "It's just empty calories." And so I (and countless dietitians) believed for many years following. My father defended his avoidance of sugar by saying the sweetener was inherently harmful to our teeth and bodies. He was a traditionalist, but, on this issue, he was way ahead of his time. My father died several years ago, at age 85, from causes other than heart disease. I now believe his choice to avoid sugar added a decade to his life.

Yudkin, Lustig, Malhotra, and Schmidt are just the tip of the iceberg of changing views about sugar. Physicians and scientists increasingly agree epidemiology and laboratory research, as well as clinical trials, clearly show sugar is not the benign substance it was once considered. And no one—not even the young, fit, and trim—is immune to its harmful effects.

Beyond risks to coronary health and the connections to diabetes and metabolic syndrome, there is also evidence sugar is connected with diminished brain function. Shaheen E. Lakhan, MD, PhD, MEd, executive director of the Global Neuroscience Initiative Foundation, believes "non-natural fructose" affects mental health, particularly cognitive decline and memory.[93] "[S]everal studies have shown intake of added sugars is associated with lower cognitive function," he says; "the association between added sugar intake and MMSE [metabolic syndrome] was independent of BMI [body mass index] and age."

A recent *Medscape* article noted that Dr. Lakhan has been at the forefront of some of the newest research on mechanisms connecting diet and mental health, specifically suggesting a relationship with the microbiota in the gut. "The emerging evidence suggests the pathway to be diet–microbiota–inflammation–mental health," he explains.[94] This science is still emerging, but current research suggests our gut flora, among the first cells to come in contact with our food, may play a key role in physical and mental health.

Dealing with the scourge of dietary sugar is a thorny public health challenge. Humans are instinctively driven to seek out sweet foods. Exceptionally rare in nature, sweetness signals something ready-to-eat, tasty, and full of energy. No known naturally sweet food is poisonous. But in nature, fruits are ripe only fleetingly, and typically come around just once a year. The sugar in fruits is tightly bound to loads of fiber, making it slow to digest and difficult to overeat. Most experts agree that the sugar occurring naturally in fruits and vegetables, including the fructose, is not the problem. Modern processed foods, loaded with sweeteners, even in savory foods, take advantage of our innate attraction to sugar.

Compounding the problem, sugar is physiologically addictive. Recent studies have documented that sugar affects the body and brain in much the same manner as substances like cocaine.[95] Like other drugs, sugar fosters a nearly irrepressible urge to consume more of it. Research at Princeton found rats can be turned into sugar junkies.[96] Clinical research on humans, done by the Oregon Research Institute in 2013, demonstrated that the brain's "reward center" is activated by sweetness much more than fat.[97] It's the sugar we crave. The more we eat, the more we crave.

Carbohydrates, especially sugar, trigger the brain to make us feel good. State of mind is closely connected to physiology. "Happiness is also a biochemical state, mediated by the neurotransmitter serotonin," explains Robert Lustig. "One way to increase serotonin synthesis in the brain is to eat lots of carbohydrates." Yet as with other addictive substances, over time our bodies require more and more of them to get the same sensation. "Eating more carbohydrates, especially sugar, initially does double duty: it facilitates serotonin transport and it substitutes pleasure for happiness in the short term. But as the D2 receptor [in the brain] down-regulates, more sugar is needed for the same effect."[98]

Here I'll pause momentarily for a personal confession: My sugar habit went well beyond high school; it has been a lifelong struggle. Research about the addictive nature of sugar aligns with my own experiences. As an adult, I spent a decade very consciously curbing my sugar—eliminating all sweetened beverages, cutting out candy almost entirely, and limiting my portions of any sweet snack or dessert. Nonetheless, I continued to have an almost uncontrollable urge to eat something sweet after every lunch and dinner. Like Pavlov's dogs hearing that bell: *Finish meal, crave sweets.* The only way I have found to rid myself entirely of the longings for sugar is completely eliminating it from my diet, something I did as an experiment for nearly a year. Research now suggests my lifelong sweet tooth is both learned and biological.

Much as I adore and admire her, my mom may bear some of the responsibility. Like clockwork, she always had a little sweet thing every afternoon and evening. Cutting-edge research suggests a mother's diet affects her baby's food preferences as it develops in the womb.[99] Some part of my sweet craving may already have been in place when I was born. As Americans have been eating ever more sugar, developing fetuses have also been exposed to ever-greater amounts. The American sweet tooth has been a self-perpetuating and intensifying cycle.

On the plus side, though, while standard medical advice of the day was, "Why bother breastfeeding?," my mom did. A study on US use of formula notes: "During the 1950s and 1960s, the trend in breast-feeding was steadily downward, and by the early 1970s, only 25 percent of infants were breast-fed at age one-week and only 14 percent between two and three months of age."[100] In his book *Fat Chance*, Dr. Robert Lustig explicitly mentions "less breastfeeding" as among the major causes of today's obesity epidemic.[101] America's switch to rearing babies on formula is likely another factor in today's chronic health problems.

Formula's cloying sweetness is part of the problem. Human breast milk contains no sucrose at all. Sucrose, remember, is a combination of fructose and glucose. Breast milk contains only lactose (7.2 percent), which is considerably less sweet.[102] In the 1950s, sucrose was the primary sweetener in baby formula, and, while it has been taken out of some formulas, it remains in major brands on US grocery shelves. Similac Organic and Similac Soy Formula each contains about 1 teaspoon of sucrose in every 5 ounces. A *New York Times* blind tasting panel of infant formulas unanimously identified those sweetened with sucrose as far sweeter than those with lactose.

Overly sweet baby formula causes two sets of problems. For one, it triggers overeating. Dr. Benjamin Caballero, director of the Center for Human Nutrition at the Johns Hopkins Bloomberg School of Public Health and an expert in risk factors for childhood obesity, is troubled by unnaturally sweet infant formulas. "[S]weet tastes tend to encourage consumption of excessive amounts," Dr. Caballero says. He points to evidence showing that babies and children will always show a preference for the sweetest food available. They will also eat more of it than they would of less-sweet food.[103]

On top of that, highly sweet formulas may start babies down the path of lifelong sugar cravings. Pediatric dentist and nutritionist Dr. Kevin Boyd says of sucrose-sweetened formulas: "We're conditioning them to crave sweetness." Formula with sucrose, which he calls "super sweet," will foster a cycle of children craving sugar, and "makes the kid want to eat more."[104] According to Dr. Ivan de Araujo, a fellow at the John B. Pierce Laboratory at Yale University School of Medicine, new research shows animals prefer sucrose over other sugars, and eating sucrose generates sugar cravings. Putting sucrose in baby formulas can have long-term negative consequences, de Araujo agrees.[105]

Although it's unclear precisely why (how much is formula composition, how much is overconsumption, how much is it the bottle as delivery mechanism?), a host of studies have shown that formula-fed babies are are more likely to end up with various health problems, including being overweight. Researchers at Brigham Young University found that compared with breast-fed babies, infants fed formula in the first six months more than twice as likely to be obese by age two.[106] Studies have also connected formula feeding with a wide range of baby health problems, including ear infections, food allergies, and asthma.[107] A study in the journal *Pediatrics* estimated that if 90 percent of US families followed guidelines to breastfeed exclusively for six months, the country would annually save $13 billion from reduced medical and other costs.[108]

Because of studies like these, the American Academy of Pediatrics recommends exclusively breastfeeding babies until the age of six months, and continuing to breastfeed after the introduction of solid foods until the baby is at least one year old. Unfortunately, the reality is quite different. Only 20 percent of US infants are nursed at all by the time they reach six months.[109] Bottle-feeding will continue to aggravate the obesity crisis in the years ahead.

Alternative Hypothesis, Part 2: Grains

While the case against sugar is clear, another group of carbohydrates may also be contributing to the rise in obesity and chronic diseases: grains.

In contrast with meat, animal fats, or sugar, the wholesomeness of wheat and other grains has rarely been questioned. Expressions like *the bread of life* indicate the central role grains have long held in the human diet. Wheat makes up 20 percent of the world's dietary calories. But some physicians and scientists have begun to question whether grain-based foods—especially from modern strains—should be a dietary staple.

The body treats all refined carbohydrates in a similar fashion. Glycemic index is a modern measure of how foods affect blood sugar and insulin. The higher a food's glycemic index, the more it affects blood sugar levels. "Foods with a high glycemic index, like white bread, are rapidly digested and cause substantial fluctuations in blood sugar," explains the Harvard School of Public Health's newsletter.[110]

Yet some believe the problem goes well beyond white flour. Dr. William Davis, a preventive cardiologist practicing in Milwaukee, Wisconsin, makes the case in his book *Wheat Belly*. Bread, he urges, is not the staff of life but a danger to human health.[111] Davis believes much of the problem originates from modern wheat. A high-yield dwarf, faster-growing plant, it's about 2.5 feet shorter than older varieties of wheat with a much larger seed. Even whole wheat is problematic. The human body rapidly converts modern wheat to sugar because it contains amylopectin A, a carbohydrate unique to wheat. Eating two slices of whole wheat bread increases your blood sugar more than a candy bar does, according to Davis. Blood sugar soon plunges and you feel hungry again, leading to weight gain and, specifically, the accumulation of visceral fat (fat around your internal organs).

Dr. Davis's studies and personal observations have led to his concern about wheat's health risks. In his practice, he started telling his pre-diabetic and diabetic patients to remove all wheat from their diets. Patients who followed his advice saw dramatic health improvements, Davis says. Eliminating wheat has led his patients to weight loss, improvements with asthma, acid reflux, mental clarity, and more, Davis reports.[112] "I've seen this with thousands of patients."

Moreover, Dr. Davis considers carbohydrates the greatest risk for formation of small low-density lipoprotein particles (LDL, the dangerous cholesterol), leading to atherosclerotic plaque, which in turn triggers heart disease and stroke. Eating carbohydrates increases heart disease risk, he says, "even if you're a slender, vigorous, healthy person."

Additionally, Davis points out, "carbohydrates increase your blood sugars, which cause this process of glycation, that is, the glucose modification of proteins." Glycation is the bonding of a protein or lipid molecule to a sugar molecule, fructose or glucose. This process creates inflammation, which can activate your immune system.[113] And glycating small LDL makes you more prone to atherosclerosis.

Davis's views of carbohydrate-heavy diets, and especially wheat, is shared by Dr. David Perlmutter, a board-certified neurologist and a fellow of the American College of Nutrition who practices in Naples, Florida. "Carbohydrates are absolutely at the cornerstone of all of our major degenerative conditions," Perlmutter states unequivocally.[114] He includes Alzheimer's, heart disease, and cancers. Elevations in blood sugar, even mild ones, compromise brain structure, lead to brain shrinkage, and "are strongly related to developing Alzheimer's disease," according to Perlmutter. Grain-based foods have a high glycemic index, cause carbohydrate surge and inflammation, and are therefore detrimental to the brain, he urges. "Inflammation is the cornerstone of Alzheimer's disease and Parkinson's, multiple sclerosis—all of the neurodegenerative diseases are really predicated on inflammation."

The human diet consists of three macronutrients: carbohydrates, protein, and fat. Americans today eat about 60 percent carbohydrate, 20 percent protein, and 20 percent fat (much of the fat from industrial seed oils). These levels contrast starkly with the optimal diet, according to Perlmutter. He believes an optimal diet is around 75 percent fat, 20 percent protein, and just 5 percent carbohydrates.

Some have accused Perlmutter of advocating an extreme, radical dietary change. But Perlmutter answers he is not suggesting change at all. Rather, he advocates getting back to a diet with which our bodies evolved. As he puts it: "I am recommending that we end this grand experiment and return to a diet that isn't evolutionarily discordant." The extreme change, he notes, is the dietary shift of recent centuries. "In the early 19th century, Americans consumed just over 6 pounds of sugar each year. That figure now exceeds 100 pounds. And there has been a dramatic reduction in the consumption of healthful fat." The consequences are protein glycation, uncontrolled insulin signaling, and unknown epigenetic consequences, Perlmutter contends.

In addition, Perlmutter believes widespread vitamin D shortages contribute to today's epidemic of chronic diseases. "Vitamin D is a powerful player in terms of brain health," he says. "Beyond strong and healthy bones, vitamin D activates more than 900 genes in human physiology, most of which are important for brain health. Low levels of vitamin D correlate with increased risk for multiple sclerosis, dementia, and

Parkinson's disease." As fewer people work outside, such as on farms and ranches, and more people spend time in front of screens, our population's vitamin D shortage has only gotten worse.

The Carbohydrate Connection

Gary Taubes, author of the bestselling books *Good Calories, Bad Calories* and *Why We Get Fat*, argues that the key problem in the modern diet is excessive carbohydrates—too much "sugar and flour." Metabolizing carbohydrates requires the body to release insulin. In the short term this leads to weight gain, and in the long term leads to insulin resistance, Taubes says.

The demographic data makes Taubes's argument especially compelling. We know sugar consumption has risen steadily and dramatically. The same thing has also happened in recent decades with grains. From 1970 to 2000, per-capita US grain consumption skyrocketed by a mind-boggling 48 percent.[115]

In addition to heart disease, other chronic diseases and conditions have also frequently been pinned on red meat. These include obesity, diabetes, and sometimes even brain degeneration. For each, sugar is now a more credible suspect. Ironically, there is also considerable evidence of just the reverse: that red meat helps people *avoid* each of these health problems.

Gary Taubes's books trace more than a century of clinical treatments for obesity and diabetes that found a single strategy most successful: limiting carbohydrates. These treatments not only permitted but even encouraged beef and beef fat. Modern low-carbohydrate diets, like Dr. Atkins's, expressly allow both without restriction. Until a decade or so ago, such diets were controversial because they ran so directly counter to the prevailing high-carbohydrate, low-fat diet that had so thoroughly captivated America's public health and medical mind-set.

"Over the past five years, however," Taubes wrote in 2002, "there has been a subtle shift in the scientific consensus. It used to be that even considering the possibility of the alternative hypothesis, let alone researching it, was tantamount to quackery by association. Now a small but growing minority of establishment researchers have come to take seriously what the low-carb-diet doctors have been saying all along."[116]

Even conservative institutions like the Harvard School of Public Health have acknowledged the shift. Posing the question in its January

2014 newsletter: "Will going on a low-carbohydrate diet help me lose weight?" it referred to "much debate about the impact of low carbohydrate eating pattern on overall health." It then cited a recent 20-year study on 82,080 women, which "found that a low-carbohydrate eating pattern did not increase risk of heart disease."[117]

Slowly but steadily, new research is transforming the scientific consensus. Many studies now document damage to the human body from overconsumption of sugar and other carbohydrates. Simultaneously, a growing body of research is documenting the efficacy of low-carbohydrate diets for weight loss and diabetes control.

One study has been especially influential. A well-funded yearlong clinical trial carried out by researchers at Stanford University School of Medicine was billed as "the largest- and longest-ever comparison of four popular diets."[118] Previous clinical trials, the researchers noted, had found that low-carbohydrate, non-energy-restricted diets "were at least as effective as low-fat, high-carbohydrate diets in inducing weight loss." Yet they were limited by small sample sizes, high rates of attrition, short durations, or limited diet assessment. For the Stanford study, more than 300 women were enrolled, each randomly assigned to one of four groups, with diets ranging from very-high-carbohydrate, low-fat diets (the "Ornish" group), to a low-carbohydrate, high-fat diet (the "Atkins" group). The other two diets had intermediate levels of fats and carbohydrates.

The press release's title suggests the study's unequivocal findings: "Stanford Diet Study Tips Scale in Favor of Atkins Low Carb." "The lowest-carbohydrate Atkins diet came out on top," the press release began, continuing: "Those randomly assigned to follow the Atkins diet for a year not only lost more weight than the other participants, but also experienced the most benefits in terms of cholesterol and blood pressure."

It's important to note the Stanford study went beyond establishing the efficacy of low-carbohydrate diets for weight loss. It also exonerated them from suspicions they are otherwise unhealthy. In an accompanying article published in the *Journal of the American Medical Association*, the researchers concluded: "We interpret these findings to suggest that there were no adverse effects on the lipid variables for women following the Atkins diet compared with the other diets and, furthermore, no adverse effects were observed on any weight-related variable measured in this study at any time point for the Atkins group."

Unlike the all-too-common scenario where experiments are carried out to "prove" what researchers want to show, the Stanford team had not initiated the study to advance low-carbohydrate diets. Quite the reverse. The lead researcher, Christopher Gardner, PhD, a longtime vegetarian, acknowledged having launched the study partly due to his concerns that Atkins-type diets might be dangerous. "Many health professionals, including us, have either dismissed the value of very-low-carbohydrate diets for weight loss or been very skeptical of them," he said. "But it seems to be a viable alternative for dieters."[119]

Concerns over negative health effects of low-carbohydrate diets, high in total and saturated fat, simply did not pan out, the researchers reported. "[R]ecent trials, like the current study, have consistently reported that triglycerides, HDL-C, blood pressure, and measures of insulin resistance either were not significantly different or were more favorable for the very-low-carbohydrate groups."

The only exception seemed to be LDL cholesterol, which went up for the low-carbohydrate dieters. But here, as well, the researchers concluded that the effect was not cause for concern. "Although a higher LDL-C concentration would appear to be an adverse effect, this may not be the case under these study conditions." They went on to explain the triglyceride-lowering effect of a low-carbohydrate diet leads to an increase in LDL particle size, which is known to decrease LDL atherogenicity. While the LDL concentrations increased slightly (2 percent), triglyceride concentrations decreased by 30 percent. The study did not assess LDL particle size, but the researchers concluded that their findings were consistent with a beneficial increase in LDL particle size. Thus, the Stanford researchers concluded, the low-carbohydrate diet was the most effective for weight loss and had no noteworthy downsides.

What About Diabetes?

A considerable amount of research links type 2 diabetes to sugar. There's a direct link to sugar, through the ways it affects the body's metabolism, and there's an indirect link, by contributing to obesity. In both ways, eating sugar raises a person's risk of developing diabetes.

Some studies have purported to show a similar link between diabetes and red meat. This claim has far less empirical support and is probably dead wrong. Being overweight is the most important factor for type 2 diabetes, according to both Mayo Clinic and the Centers for Disease Control and Prevention. Yet as the Stanford study showed, red meat and fat help people lose and control their weight.

Accurately assessing the alleged link between red meat and diabetes is challenging. One problem is the healthy user bias discussed earlier. People motivated by health and fitness have heard anti-beef dietary advice for many years. That has had a notable impact on which people regularly consume beef. People who eat less red meat and saturated fat also tend to eat less refined sugar, salt, trans fats, and processed foods. The group that avoids red meat is also less likely to be overweight, is more physically active, and is far less likely to smoke. People likely to develop diabetes, in other words, are often the same people who do not follow a healthy lifestyle.

Also important, as mentioned earlier, most studies on meat eating fail to separate processed and unprocessed meat. This is a serious deficiency with respect to diabetes. Studies that do make the distinction have found an association with health problems—including diabetes—only from processed meats. This suggests that something other than the meat itself (perhaps salt, nitrites, or other additives) is responsible for the apparent relationship between meat and type 2 diabetes.[120]

The most important study to date that seemed to support the connection was done in 2013 by researchers from various institutions, including Harvard's School of Public Health. But that study provided only weak evidence of any link.[121] Like the studies claiming to show a connection to heart disease, it was an observational study in which researchers merely examined data, rather than a clinical trial. These studies inherently have limited value. And for this particular study, the results were inconclusive.

As explained by Chris Kresser in his article "Does Eating Red Meat Increase the Risk of Diabetes?," all but one of the associations between red meat and diabetes disappeared after researchers adjusted for body mass index (BMI). "[E]xcess body fat is the biggest risk factor for type 2 diabetes, so it shouldn't come as a surprise that people with higher BMIs have less-than-ideal biomarkers for glucose metabolism," Kresser points out. "Additionally, it's common for people who are overweight or obese to have underlying chronic inflammation, so it makes sense that people with higher BMIs would tend towards having higher levels of inflammatory biomarkers."[122]

Mark Sisson, "primal diet" advocate and author, commenting on the same diabetes study, notes it was heavily impacted by the healthy user bias. He points out not all important dietary and lifestyle factors were given consideration by the researchers (most notably, sugar consumption levels). Even with that failure, the study's data shows clearly the "heavy meat eaters" of the study were people with much less health-conscious diets and lifestyles.

> *Folks in the highest quintiles of meat intake were the least active and the most sedentary. They exercised the least and smoked the most tobacco. They drank more alcohol than any other quintile. They guzzled more soda and other sweetened beverages. In the high meat quintiles, folks ate 800 more calories per day than folks in the low meat quintiles. They were much heavier, too . . . Trans fat intake was higher in the high-meat quintiles, too, as was potato intake . . . They ate the least amount of fiber from grains, indicating they probably ate the most refined grains, drank the most coffee, and ate the fewest fruits and vegetables. In short, people who ate the most red and processed meat were also the unhealthiest by both Primal and mainstream standards . . . These people (all health professionals, ironically) most likely didn't particularly care about their health . . . Added sugar . . . wasn't [even] covered.*[123]

Moreover, as Gary Taubes argues compellingly in his books, the greatest cause of obesity today is excessive carbohydrate consumption. Beef contains no carbohydrates. It's rich in high-quality protein and healthy fats. And as the Stanford study and a wealth of other research has demonstrated, beef is a valuable part of a weight-controlling diet.

A Novel Theory: TMAO

In the past few years, a new theory about red meat has surfaced. It has to do with the way it may affect the microbes living in our digestive tracts. Researchers from the Cleveland Clinic hypothesized in 2013 that eating red meat predisposes the gut flora toward production of trimethylamine N-oxide (TMAO), which has been associated with increased heart attack risk.[124] The clinic's news release also stated the same level of TMAO production did not occur in the body of a vegan, even when she ate meat.

The red meat–TMAO theory is a novel idea with very little credible science to back it up. Commenting on the Cleveland Clinic study, cardiologist Mozaffarian stated it was simply "too early" to know whether the TMAO molecule causes atherosclerosis in humans.[125] Recent studies have shown excluding meat from your diet does not lead to greater longevity or better overall health,[126] which undermines the credibility of the TMAO hypothesis. It also fails to explain why major epidemiological studies, like the ones cited previously, have shown no relationship between red meat consumption and heart disease.

Moreover, other differences in diet or behavior may actually explain the higher presence of TMAO in the blood of meat eaters. For example, the red meat eaters involved in this study may also have eaten fewer fruits and vegetables and more refined carbohydrates, flour, and sugar. Previous studies have shown such eating habits to be more prevalent among red meat eaters, and each of them has been associated with undesirable changes in the gut microbiota.[127] Earlier studies by the same researchers also showed that people with higher levels of certain gut bacteria produce higher levels of TMAO. Research by others has shown that the TMAO-producing gut bacteria is associated with consumption of whole grains.[128]

More recently, research published in the journal *Nutrients* in October 2018 takes careful inventory of all TMAO research and theories. The article vividly illustrates the vastness of the uncertainty—starting with the most fundamental question of whether TMAO actually causes problems or is merely a biomarker. It also notes that blood levels of TMAO are determined by several factors. Foods found to increase TMAO include fish, mushrooms, and beans. Vegetables, like those of the Brassicaceae family (such as kale and broccoli), affect TMAO formation in uncertain ways. Beyond diet, factors that affect TMAO formation include gut microflora, various pharmaceuticals, and liver flavin monooxygenase activity. Elevated TMAO levels and atherosclerosis are correlated, the study notes but "the exact mechanism underlying this correlation is still unknown."[129]

There are so many unanswered questions. What makes the proposed chain of events with red meat, TMAO, and disease highly improbable to me is we humans have been eating red meat for millions of years. It's just not logical that a food upon which our ancestors depended and thrived is inherently dangerous. The only thing certain about the TMAO hypothesis is it's far from a credible indictment of beef.

Is Red Meat Linked with Cancer?

Similar questions have clouded inquiries into a purported link between red meat and cancer. The research is fraught with problems. For one thing, "[d]espite numerous investigations, to date none of these hypotheses have been able to convincingly explain the link between red meat intake and cancer risk," notes a 2013 *Medscape* article.[130] No credible explanation for a physiological mechanism has panned out in the research.

Additionally, like those about heart disease, studies on red meat and cancer have undoubtedly been heavily influenced by the healthy user bias. This bias may, in fact, be strongest in cancer studies. A lot of research suggests that antioxidants found in fruits, vegetables, and fresh greens assist the human body in defending itself against cancer. Health-conscious people often avoid red meat and eat copious amounts of fruits and vegetables. So the group eating the most fresh produce overlaps considerably with the population consuming little or no beef. This makes conclusions about beef's supposed connection to cancer all the more suspect.

Studies of meat's possible connection to cancer also fail to make crucial distinctions. Most do not consider the form of meat—fresh versus processed. Nearly all fail to consider the means of preparation—grilled, high heat, or braised. None consider the farming systems employed. Yet, as noted, these are essential differences. These factors determine healthfulness of meat when it's consumed, particularly with respect to its cancer risk. For instance, a major European study in 2013 in which a large cohort (almost half a million individuals) was followed for a median of 12.7 years found a higher cancer risk for processed meats (like sausages and bacon) but no increased risk for unprocessed meats.[131]

The type of cancer given the most media attention for its purported connection to meat is colon cancer. A 2011 study in the *European Journal of Cancer Prevention*, "Meta-Analysis of Prospective Studies of Red Meat

Consumption and Colorectal Cancer," sought to address the considerable "scientific debate" about the association between red meat and colorectal cancer (CRC). The authors conducted a meta-analysis of 25 independent, non-overlapping prospective studies. Researchers found the associations "weak" and "inconsistent," and found "the likely influence of confounding by other dietary and lifestyle factors." They conclude: "The available epidemiologic data are not sufficient to support an independent and unequivocal positive association between red meat intake and CRC."[132]

In addition to colon cancer, a possible association between red meat and breast cancer has garnered considerable attention. Yet a major study by researchers from the Harvard School of Public Health in 2002 found no credible link between breast cancer and milk or meat. Researchers sought to assess "the risk of breast cancer associated with meat and dairy food consumption and . . . whether non-dietary risk factors modify the relation." More than 20 studies, they noted, had previously investigated the issue with conflicting results. The Harvard team combined primary data from eight prospective cohort studies from North America and Western Europe with at least 200 incident breast cancer cases, assessment of usual food and nutrient intakes, and a validation study of the dietary assessment instrument. The data included 351,041 women, 7,379 of whom were diagnosed with invasive breast cancer during up to 15 years of follow-up. The team's investigation concluded: "We found no significant associations between intake of meat or dairy products and risk of breast cancer."[133]

This Harvard review shows studies of dairy products and breast cancer have conflicting results. As with red meat, recent, carefully executed research has shown no convincing evidence that dairy causes breast cancer. Another study reinforcing this was published in the *International Journal of Fertility and Women's Medicine* in 2005. Its aim was "to review the current epidemiological literature on the relationship between the intake of dairy products and breast cancer risk." Researchers reviewed 39 case-control studies, 11 cohort studies, two meta-analyses, and several review articles. Looking at this large body of evidence, the team concluded: "There is no substantial epidemiological evidence to support a significant link between the intake of dairy products and breast cancer risk."[134]

Some studies have even shown that dairy products may lower cancer risks. This is actually not particularly surprising because, like fruits and vegetables, dairy products contain a substance believed to help the body combat cancer. In dairy that compound is conjugated linoleic acid (CLA). A 2017 article in the *International Journal of Cancer Management* summarized the research as follows: "CLA affects different stages of cancer; CLA inhibits proliferation and can induce apoptosis [programmed cell death] in breast, prostate, colon, gastric, liver, endometrial and osteosarcomas cancer cells and also prevent metastasis. Furthermore, a study on mice showed that at least 0.5% CLA in the daily diet can reduce incidence of cancer. It can be concluded that CLA might be effective to control and prevent cancer."[135] CLAs are found in significant amounts in milk, and especially cheese, from cows living on grass.[136]

Other recent studies also have found no noteworthy connection between red meat and dairy for ovarian cancer[137] and pancreatic cancer.[138]

The most important thing to realize about the conflicting health research is this: The benefits of eating nutrient-rich real foods—including beef and dairy products—are clear and well documented; the risks, on the other hand, are tiny, uncertain, and unproven. This point was driven home by an important set of studies released in late 2019, which were based on three years of work by a group of 14 researchers in seven countries. In three studies, the group examined extant research regarding red meat or processed meats and the risk of cardiovascular disease or cancer. The researchers found there were no statistically meaningful health risks forming the basis for dietary guidelines that advise reducing red meat. An article on the findings in *The New York Times* said, "in a remarkable turnabout, an international collaboration of researchers produced a series of analyses concluding that the advice [to reduce red meat], a bedrock of almost all dietary guidelines, is not backed by good scientific evidence."

More specifically, the group reviewed 61 articles reporting on 55 populations, with more than four million participants. The researchers also reviewed randomized trials linking red meat to cancer and heart disease, and 73 articles that examined links between red meat and cancer incidence and mortality, the *Times* noted. "In each study, the scientists concluded that the links between eating red meat and disease and death were small, and the quality of the evidence was low to very low." Dr. Bradley Johnston, a lead researcher in the project and epidemiologist at Dalhousie University in Canada, explained that the group concluded dietary recommendations to lower red meat were not scientifically valid. The studies suggest the tangible benefits of eating red meat are not outweighed by the small and uncertain risks they may pose. "The certainty of evidence for these risk reductions was low to very low," Dr. Johnson explained.[139]

As is par for the course, this international collaboration of scientists was vehemently attacked soon after publishing these findings. There was no serious objection to either their methodologies or their analysis. Instead, vegan groups and some mainstream nutritionists accused them of bias. But all these researchers really did was look at the question scientifically and state the obvious: The association between unprocessed red meats and heart disease and cancer is and has always been feeble.

Undoubtedly, multiple, diverse circumstances conspire to create our modern epidemic of chronic diseases. Some possible factors—like antibiotics overuse and genetically modified crops—are just coming under scrutiny. Between 85 and 95 percent of several major crops are now genetically modified. Some are foods directly eaten by humans—papayas, sugar beets, potatoes, and canola, to name a few. Some, like corn and soy, are fed to animals from which human food is derived. The long-term health and environmental consequences of the genetically modified ingredients in our food chain are both unknown and virtually unknowable.

But one thing is becoming increasingly clear: The surge in chronic diseases is not due to real foods we have been eating forever like beef or bovine fat. Chronic diseases have been rising at the very moment in history that our consumption of beef and animal fat has been falling. As cardiologist Dr. Mozaffarian said in his talk "What Is the Optimal Diet to Prevent Cardiovascular Disease?," neither saturated fat nor red meat is much of a risk factor.[140] Instead, as Mozaffarian notes, evidence now strongly points to processed foods, especially those that are highly processed.

Nutrition experts around the world are particularly concerned about a category of foods they are calling "ultra-processed." These are commonly defined as "industrially manufactured substances composed of some mix of oils, fats, sugars, starch, and proteins that contain little if any whole or natural foods." These foods generally also include artificial ingredients like flavorings, colorings, emulsifiers, preservatives, and other additives to improve appearance and texture and to extend shelf life. New research suggest that these ultra-processed foods may be harming our bodies at the cellular level. Telomeres are sections of the chromosome that act to protect it and assist in DNA replication.[141] Recent research published in the *American Journal of Clinical Nutrition* suggests that telomere length (TL) is a marker of biological age that may be affected by dietary factors through oxidation and inflammation mechanisms. Study "participants with the highest UPF [ultra-processed food] consumption had almost twice the odds of having short telomeres compared with those with the lowest consumption." This is a very new area of research, and much remains to be learned. But it strengthens my own belief that the core problem in the diets of industrialized nations is processed foods.[142]

Faux Meats

I also want to take a moment here to say a word about a particular type of highly processed foods—faux meat. I've eaten plenty of veggie burgers in my day. But I worry a lot about the new generation of high-tech meat replacements holding themselves out as more environmentally sound and healthier than beef. Silicon Valley–backed "clean meat" is now everywhere. Our guilt over eating meat, we are assured, can be assuaged by consuming nuggets and patties made from plants. This may tempt some, but faux meats are not progress, they are a distraction.

The phrase *clean meat* is itself revealing. It was first used by high-tech entrepreneurs wanting to "disrupt" the food system status quo. Multinational corporations are now putting on the "clean" mantle as they rush to introduce meat-free products while attempting to distance themselves from their own troubled legacies. The US's largest confinement pork producer, Smithfield, a company I have personally litigated against in federal court over water and air pollution, recently created a vegan brand it calls Pure Farmland.

To me, this all illustrates how wrong faux meat is as a strategy for fixing the food system. I've always considered farming a profession practiced with dust on your jeans, hay in your hair, and dirt under your fingernails. At its best, farming is a daily communion with nature that brings the deeply satisfying accomplishment of nourishing people with nutrient-rich, wholesome food.

Plant patty manufacturers want us to believe we can derive the nutritional benefits and gustatory pleasures of meat without the dirty or messy parts. They imply there's no water or air pollution, no blood. Most important, by choosing their "clean" option you can avoid the moral taint of consuming animal flesh, for which there is no longer a need, they suggest. Modern food technology, it's implicit, enables us to advance beyond physiology and the brutal ways of our ancestors.

But is this really the choice we are facing? And do faux meats live up to this promise? Let's start with health.

Ethan Brown, founder of Beyond Meat, is convinced his formulation offers a more perfect food than nature provides. His spokesperson has said, "Improving human health is a core tenet of our brand." In response to such assertions, a flurry of articles has evaluated faux meat's health claims. Nearly all have concluded that this new generation of meat substitutes cannot legitimately be called healthy. Most have focused on the saturated fat content (about the same as red meat), the sodium (very high), or the genetically modified ingredients.

But my concern is more fundamental. It's about what *kinds* of foods we are feeding ourselves. Like mothers the world over, every day I strive to feed my family and myself tasty, health-supporting, *real food*. Much of my day is spent in the kitchen serving meals, packing lunches, cooking, baking, and canning. Knowing where our food comes from and how it is raised is a priority. The staples of our diet are fruits, vegetables, greens, and herbs from our garden and local farms. We eat grass-fed beef and pasture-raised turkey from our ranch, organic eggs from hens in our neighbor's backyard, fish, oysters, and crabs from local fishermen that we personally know. Our purchasing and bartering deepen friendships and strengthen our local food economy. In stark contrast with this system of transparency, trust, and relationships, high-tech faux meats are merely the latest iteration of industrial food. To me, they are utterly unappetizing and unappealing.

There is no question these products should be labeled "processed," and a number of nutrition experts call them "ultra-processed." The best-known faux meat companies in the United States have genetically modified components and over 20 total ingredients. "Their proteins are isolates, extracted mechanically from whole soy and peas. Their fats are industrial vegetable and seed oils," notes a recent article in *The New*

Republic. The article goes on to suggest these novel meat alternatives may even deserve their own—uber-processed—category:

> *In fact, companies like Impossible and Beyond have arguably created a new, higher tier of ultra-processed food. As Engadget noted, "A Cheeto or Twinkie is unambiguously synthetic." But these fake-meat products are engineered, specifically, to fool our senses into thinking they're whole foods—and then marketed, by meat companies, to change our language to reflect the trick.*[143]

Ultra-processed foods, collectively, are coming to be regarded as the single biggest dietary problem in the developed world. *The New Republic* further points out that earlier this year, "the National Institutes of Health released a landmark study showing that America's obesity epidemic is driven primarily by ultra-processed foods"—the same month that "two large European studies linked ultra-processed food consumption to cardiovascular disease and death."

Even more compelling than this sort of research is common sense. At some level, we all know our bodies have evolved with certain kinds of foods and have not evolved with engineered edible products. In his book *Nourishment*, animal nutrition expert Fred Provenza argues that the foundation of a healthy human population is "nutritional wisdom," where individuals develop the capacity to recognize foods that satisfy their particular dietary needs. Provenza argues diets should be based on bio-chemically rich foods, connected to our landscapes and cultures, taught by mothers and elders to the young, for which we develop "flavor-feedback relationships." Faux meats fail this in every respect. When such links are broken, "entire ecosystems and cultures become dysfunctional," Provenza writes. Equating *clean* and *pure* (another word used to describe faux meats) with *good* for food and farming is not just nonsensical, it's wrong.

Faux meat manufacturers also boldly promote their products as planetary saviors. Pat Brown, founder of Impossible Foods, told *The New York Times* in 2019, "Every aspect of the animal-based food industry is vastly more environmentally disruptive and resource-inefficient than any plant-based system," as he pronounced, yet again, his aim of eliminating animals entirely from the global food supply by 2035. The company's website claims that its products have "a tiny fraction of the environmental impact" of meat, using far less land, requiring less water, and emitting much lower levels of greenhouse gases. The company often compares its products with mainstream meat production. But in 2019 it grabbed media attention by directly attacking regenerative livestock farming as the "clean coal of meat." The implication was that meat is inherently environmentally destructive no matter how or where livestock are raised.

Many of us in the regenerative agriculture community responded. We noted that a leading cause of US water pollution for the past two decades has been large-scale, chemical-intensive GMO monocrop production of exactly the type from which Impossible and other faux meat companies source their ingredients. And we pointed out that ecologically optimal farming involves animals. Getting rid of cattle is throwing the baby out with the bathwater. The best models of regenerative farming demonstrate that animals are indispensable.

Will Harris of White Oak Pastures in Bluffton, Georgia, was particularly incensed by the Impossible Foods attack. His wonderful farm, which I have had the distinct pleasure of visiting, is a remarkable example of agricultural, ecological, and economic diversification. Each year, he and his family have introduced new layers of complexity to their farming operation and new foods and household products. A recent independent environmental audit at White Oak Pastures concluded that its beef operation sequesters more carbon in the soils than it emits. Thus, White Oak is being modest in referring to its beef as "carbon neutral." They have created an elegant real-world illustration of my favorite slogan, "It's not the *cow*, it's the *how*."

I believe mainstream meat production needs improvement. But this is precisely why faux meats are such a distraction. They get people like Bill Gates and other techies and their investors all excited. This siphons the energy, media attention, and money away from the work that really needs to be done to fix agriculture. We need an all-hands-on-deck effort to transform mainstream agriculture from extractive to regenerative. Ironically, true disruption of industrialized agriculture isn't coming from people like Bill Gates. It's coming from people like Will Harris and North Dakota farmer Gabe Brown, who eloquently documents his journey from conventional to regenerative farmer in the book *Dirt to Soil*.

Special mention should be made on lab-grown meats, the techiest of the high-tech faux meats. Despite investor giddiness, the cost and practicality of creating lab conditions to culture meat is in itself a serious problem that may never be overcome. But the "yuck factor" is even more grave. An in-depth industry profile in *Wired* magazine in 2018 noted the sector has a "messy problem." "None of the major players have managed to grow meat without using animal serum—a blend of growth-inducing proteins usually made from the blood of animals. The most popular is fetal bovine serum (FBS), a mixture harvested from the blood of foetuses excised from pregnant cows slaughtered in the dairy or meat industries."[144] Lab meat is now creating a high demand for this serum. Recently, I heard that some US slaughterhouses are offering incentives for farmers and ranchers to bring cull cows to slaughter pregnant. This makes my heart hurt and stomach ache.

Twice in recent years, I was invited to share a public stage to debate the merits of faux meat with Impossible's Pat Brown. To prepare, I watched some of his interviews. Most telling to me was Brown's admission that he gleans little pleasure from eating. Food, he said, is just fuel. This is kind of sad, actually. For me, food is so much more than fuel. It is a fundamental sensual pleasure. It is the foundation of vibrant health. It connects us to our families, communities, cultures, and histories.

While perhaps soothing worries about health and the environment, faux meats are definitely not the answer. They won't make us any healthier and they won't make us happier. They will only further entrench the industrialized food system at the core of our ecological and human woes. We need a much more radical and real solution that gets to the root of how we farm and how we eat. Regenerative farming with livestock that looks like nature and produces nutrient-rich foods is that solution. And it's the only one that will work.

Even though worsening US health has *not* been accompanied by rising consumption of red meat and animal fat (the reverse is true), the notion persists those foods cause health problems. The idea is still widely accepted by the general population and in the public health and medical community. But there's more than one way to prove or disprove a hypothesis. Moving beyond the demographic, epidemiologic, and clinical data, then, let's approach this question from another angle.

Health and Diets of Hunter-Gatherers

An alternative way to evaluate the impact of red meat and animal fat on human health is by considering history and anthropology. A simple but accurate way of assessing cause and effect is to examine the physical condition of groups of people, past and present, who ate (or eat) copious amounts of meat and animal fat. If those foods truly *inherently* cause heart disease and other health problems, we should see a high prevalence of affliction among those groups. If factors other than animal fats and red meat are actually to blame, the association would likely be absent.

Since I point to sugar, flour, and industrial oils as the true culprits in chronic diseases, ideally we would examine populations who eat red meat but do *not* eat those foods. Unfortunately, sugar, flour, and seed oils have been widely distributed for over a century, making this somewhat difficult. But it's not impossible.

Historians and anthropologists have studied hunter-gatherer diets around the globe. All of these societies eat animal-based foods: flesh, fat, organs, blood, milk, and eggs. They cultivate no grains, sugar, or seed oil crops. Until the 20th century, many also had no access through trade to any of those industrial foods, and some still use them sparingly or not at all.

Anthropology professors Michael Gurven and Hillard Kaplan examined hunter-gatherer health around the world.[145] They wanted to know whether hunter-gatherers were afflicted by chronic diseases and what was their typical life span. Gurven and Kaplan found infant mortality and death by accident and injury common. On the other hand, they found chronic diseases rare. Just 9 percent of deaths were caused by chronic diseases. By contrast, in the United States chronic diseases cause two-thirds of all deaths.[146]

Degenerative deaths are relatively few, confined largely to problems early in infancy and late-age cerebrovascular problems, as well as attributions of "old age" in the absence of obvious symptoms or pathology. Heart attacks and strokes appear rare and do not account for these old-age deaths, which tend to occur when sleeping . . . Obesity is rare, hypertension is low, cholesterol and triglyceride levels are low, and maximal oxygen uptake is high. Overall, degenerative disease accounts for 6–24 percent (average 9 percent) of deaths . . .[147]

As hunter-gatherers, humans were better nourished and, consequently, healthier. Writing from the perspective of a practicing dentist, Dr. Steven Lin blames our diet for most of our dental maladies. In his book *The Dental Diet*, Lin connects modern, industrial diets to poor dental health and overall health problems. Tooth decay and gum disease from our high-sugar, high-carbohydrate diet is clear. But Lin goes well beyond cavities. Citing the work of anthropologists and other researchers, he argues that the very shapes of our faces—our cheekbones, dental arches, and jaws—are malformed due to our poor nutrition. Well-nourished people around the world, especially those following traditional diets, Lin explains, have well-formed, angular faces and plenty of room for their teeth.

> *Anthropologist Robert Corruccini has looked at thousands of modern and ancient human jaws and teeth. He's also studied contemporary urban and rural populations in Kentucky and found that dental crowding was linked to the adoption of the modern diet over the traditional one. Corruccini attributes malocclusion [misaligned teeth] directly to the consumption of soft, processed foods, which are much more prevalent in modern diets. He calls malocclusion "the malady of civilization."*[148]

When humans began to depend on farmed crops their health suffered, notes Harold McGee in his treatise *On Food and Cooking*. "Thanks to the combination of meat, calcium-rich leaf foods, and a vigorous life, the early hunter-gatherers were robust, with strong skeletons, jaws, and teeth."[149] With settlement and the advent of agriculture, human health steadily declined. But late in the 19th century, there was a "return to something like the robustness of the hunter-gatherers," according to McGee. This was because more people were once again getting access to animal-derived foods. "[T]he growing nutritional contribution of meat and milk . . . played an essential role."[150]

It is sometimes suggested that chronic diseases fail to manifest in hunter-gatherer communities simply because those people die young. Some earlier researchers suggested hunter-gatherers' lives were "nasty, brutish, and short," rarely longer than 40 or 50 years. However, Gurven and Kaplan's study found no truth in that claim. Instead, for individuals who reach maturity, it was not uncommon for them to live 68 to 78 years.

Among hunter-gatherer groups whose diets and health have been formally studied are the Inuit, other Native Americans, and the Masai.[151]

Native American diets varied by climate and locale. Before the start of agriculture, all were heavily based on the flesh and fat of animals. This included, depending on season and location, "deer, buffalo, wild sheep, goat, antelope, moose, elk, caribou, bear and peccary," as well as small animals, such as "beaver, rabbit, squirrel, skunk, muskrat and raccoon; reptiles including snakes, lizards, turtles, and alligators; fish and shellfish; wild birds including ducks and geese; and . . . sea mammals."

Many people assume Native Americans ate only lean wild animals, and that, correspondingly, they consumed little animal fat and negligible quantities of saturated fat. But zoological information about the animals they ate tells a different story. Other than seals and squirrels, "the fat of all the other animals that the Indians hunted and ate contained less than 10 percent polyunsaturated fatty acids, some less than 2 percent. Most prized was the internal kidney fat of ruminant animals, which can be as high as 65 percent saturated."[152]

Historical accounts reveal the value traditional Native Americans placed on animal fats. Anthropologist and Arctic explorer Vilhjalmur Stefansson, who reported in detail about the years he spent living among northern Native Americans, noted they preferred "the flesh of older animals" rather than calves, yearlings, or two-year-olds. Older animals were prized for their thick slab of fat along the back. "In an animal of 1,000 pounds, this slab could weigh 40 to 50 pounds. Another 20 to 30 pounds of highly saturated fat could be removed from the cavity." All this fat was saved, some rendered, and used to flavor and preserve foods. Depending on the season, animal fat contributed as much as 80 percent of total calories in the diets of the northern Indians.[153] Remember, this is just about the amount of fat Dr. Perlmutter considers optimal in the human diet.

The diet of the Inuit, native people of Arctic Canada, Greenland, and the United States, is also instructive. The traditional Inuit diet was heavy on red meat and fat. It was also nearly devoid of plants. Staples included seals, whales, walruses, moose, caribou, reindeer, ducks, geese, quail, salmon, whitefish, tomcod, pike, char, and crabs. Seal oil was used for cooking and as a sauce.[154] "[T]he traditional Eskimo diet had little in the way of plant food, no agricultural or dairy products, and was unusually low in carbohydrates." Modern analyses of Inuit health are confounded by various factors, including astronomically high smoking rates (79 percent). Nonetheless, studies show that Inuit people whose diets are the closest to the traditional diet—those consuming large amounts of animal flesh and fats, including saturated fat—are the least likely to succumb to heart disease.

There is limited written record on the health of early Native Americans. What does exist suggests their health was generally excellent and nearly free of chronic diseases. Early explorers, including 16th-century Spaniard Cabeza de Vaca, wrote admiringly of the exceptional vigor of Native Americans. "The men could run after a deer for an entire day without resting and without apparent fatigue," he said. Dr. Weston Price, a dentist who, in the 1930s, spent considerable time among native communities in various parts of the world, "noted an almost complete absence of tooth decay and dental deformities among native Americans who lived as their ancestors did."[155] Price also interviewed health care practitioners, including a Dr. Romig in Alaska, who reported "in his thirty-six years of contact with [native] people he had never seen a case of malignant disease among the truly primitive Eskimos and Indians, although it frequently occurs when they become modernized."

Complementing such accounts are archaeological investigations, particularly examinations of human bones and teeth. One line of this research has uncovered the major chronic health problem in ancient New World populations due to iron-deficiency anemia. But the evidence also suggests that the anemia came from abandoning the hunter-gatherer lifestyle and increasingly relying on agriculture, specifically maize cultivation. "[I]ncreased population density [from the development of agriculture in North America], along with other ecologic and demographic changes associated with intensified farming, had a profound influence on health, with statistically significant increases in infectious diseases and iron deficiency anemia."[156]

Looking beyond the Americas, the Masai people have long fascinated Westerners. A livestock-herding people living in Kenya and Tanzania, the Masai are living contradictions to the prevailing anti-animal-fat, anti-red-meat dogma of industrialized countries. "Their diet," a 2012 *Science Nordic* article notes, "is as full of fats as the diet of people living in the

West."[157] But their fats come from cattle, not industrial oils manufactured from soy, corn, palm, and canola. Historical accounts by Moritz Merker, a German who lived among the Masai for eight years, beginning in 1895, suggest the traditional Masai diet consisted mostly of dairy and red meat. "The favorite foods of the Masai, Merker tells us, were milk, meat, and blood. They would drink the blood of cattle, sheep, or goats in the fresh or clotted state."[158] "Unlike Westerners, however, the Masai do not have many problems related to lifestyle diseases."[159] Newer research shows that counter to some previous claims, the good health of the Masai is not due to intense exercise. Instead, the Masai engage in "moderate but constant physical activity." Regardless, the lack of chronic diseases among the Masai shows there is nothing inherent in red meat, animal fats, and dairy that causes heart disease or other chronic diseases.

The diets of many traditional cultures were or are rich in red meat and fats, including plenty of saturated fat. Yet none had or have high rates of diet-related diseases or conditions. Clearly, nothing contained in those real foods causes chronic diseases.

The extensive investigative work of Dr. Weston Price, especially, documented excellent health among people following traditional diets. "Dr. Price found fourteen groups—from isolated Irish and Swiss, from Eskimos to Africans—in which almost every member of the tribe or village enjoyed superb health. They were free of chronic diseases . . ."[160]

Everywhere in the world, health declined as diets became more Westernized—in other words, more processed. In *Guns, Germs, and Steel*, Jared Diamond points out the common paradox that as human societies became more "technologically advanced" they also became less well nourished. "Food production, while increasing the quantity of edible calories per acre, left the food producers less well-nourished than the hunter-gatherers whom they succeeded."[161]

Cheap Food Is Cheap

Sugar, flour, and seed oils. These are the cornerstones of the modern, industrial diet. We've made these processed food ingredients inexpensive and plentiful. We overconsume them. Our collective health is paying the price. Our current culture of excess no longer regards food as precious. We expect it to be cheap, and it is.

Not long ago we Americans spent much more to nourish ourselves. In 1933, 25 percent of our income went to food. By 1950 this was up to 31 percent. But today we spend just over 9 percent of our income on food.[162] How much we spend on food runs parallel with our attitudes: What we obtain cheaply, we expect little value from.

For decades, America has massively overproduced grain and other commodities. People think factory farms and feedlots became necessary to feed a growing population. But this belies the historical facts. In reality, American farms have long overproduced. In their treatise on agriculture, *American Farm Policy 1948–1973*, professor of agriculture Willard Cochrane and agricultural research fellow Mary Ryan conclude: "Except in wartime there have been too many resources producing too much product—excess production capacity—on American farms since 1930; this is the basic problem of the commercial farm sector."

At the same time, we throw away more food than ever. Nearly half of all food produced in the United States is never eaten. By weight, food waste is the single largest component of American garbage. Recovery of just 25 percent of wasted food would feed 20 million people, USDA estimates.

We exacerbate the crises by propping up production of unhealthy foods and additives. Sugar, nutritionally the least valuable and among the most hazardous, is subsidized by American taxpayers to the tune of $4 billion per year. Corn (used both as a grain and to make high-fructose corn syrup) has been subsidized at $80 billion over the past 15 years. Around 45 percent of farm subsidies are funneled into large scale mono-crops used for feeds, especially soy and corn. An additional 30 percent goes to wheat and rice. These crops are also the main components of the processed foods Americans are consuming in excess.

Our current industrialized food system encourages overproduction, overeating, and waste. The result is environmental damage and an unhealthy population. With cheap food we are getting what we pay for. As Michael Pollan has brilliantly quipped, "You pay your grocer now or pay your doctor later."

CHAPTER 8

Beef Is Good Food

HOPEFULLY BY THIS POINT YOU'LL AGREE BEEF HAS BEEN WRONG-fully accused of unhealthfulness. But you may still wonder, why eat it? We've been deeply trained to mistrust both the environmental role of cattle and the healthfulness of beef. Perhaps you've gotten in the habit of avoiding it or eating it only rarely?

Even so, most people are, at least some of the time, strongly drawn to a grilled hamburger or steak, a roast beef or pastrami sandwich, or spaghetti with meatballs. Other than brewing coffee, or perhaps frying bacon, I find few smells more tantalizing than beef on a charcoal grill. Just a few days before writing this, a longtime vegetarian friend confessed on Facebook she'd caved and eaten a beef burger the night before. What pulls us toward beef despite being been told so often we shouldn't eat it?

In mulling over the question, "Why do people love meat?" food scientist Harold McGee writes:

> [T]he deepest satisfaction in meat probably comes from instinct and biology. Before we became creatures of culture, nutritional wisdom was built into our sensory system, our taste buds, odor receptors, and brain. Our taste buds in particular are designed to help us recognize and pursue important nutrients: we have receptors for essential salts, for energy-rich sugars, for amino acids, the building blocks of proteins, for energy-bearing molecules called nucleotides. Raw meat triggers all these tastes, because muscle cells are relatively fragile, and because they're biochemically very active . . . Meat is thus mouth-filling in a way that few plant foods are. Its rich aroma when cooked comes from the same biochemical complexity.[1]

Our instincts draw us toward beef. At the same time, our brains—ringing with alarm bells and warning whistles crammed in there by modern health and ecological advice—may turn us away. Yet I am increasingly convinced we need to let those instincts guide us more in life, especially when it comes to food. That deeply inborn impulse we so often ignore or suppress is often leading us in the right direction. Our bodies know that the milk, fat, broth, and flesh of cattle are delicious and nutrient-dense. As Dr. Fred Provenza teaches in his book *Nourishment*, we—each of us—hold innate nutritional wisdom. We intuitively know those foods that are uniquely able to provide the human body nourishment in a usable form. "[A]n impartial review of the evidence indicates that red meat is one of the healthiest foods you can eat," writes nutrition expert Chris Kresser.[2]

Theories abound about what we ought to eat and why. We are told everything from that we should never eat meat to that an all-meat diet is best. For my part, I find the following straightforward idea most credible:

We should eat what our bodies evolved to eat.

Ancestors of modern humans began consuming animals at least 2.6 million years ago. They started eating substantial amounts around 1.5 million years ago.[3] Our brains increased in size both because of meat's rich nourishment and because of hunting's inherent complexities. Over millions of years, eating meat has been interwoven into the very fabric of our evolution.

Being omnivores has been a tremendous survival advantage. On the opposite end of the spectrum are koalas, pandas, and monarch butterflies. Each can survive only on a narrow range of foods. If those specific food sources are wiped out, their extinction is assured. We humans, by contrast, are highly adaptable. We can thrive on tens of thousands of plants, animals, and fungi. Our diets have always varied widely, closely reflecting the climate and geography in which we reside. Every group of humans around the globe has always eaten from a smorgasbord of local offerings that differs considerably from those of other regions.

I do not believe in any one, hard-and-fast dietary dogma (in fact, I find most rigid diets a bit silly). But I think it's clear our bodies benefit greatly from eating the animals of the field, as our ancestors before us did for countless generations. Bison, antelope, moose, alpacas, yak, gazelles, deer, elk, camels, and caribou are all grazing animals uniquely able to live from grass and other cellulosic plants on which humans themselves cannot survive. So are cattle. Like other grazing animals, cattle, symbiotically with microorganisms, miraculously convert those plants into milk and flesh, which are jam-packed full of essential nutrients. And in contrast with those found in plants, nutrients in bovine milk and meat are uniquely usable by the human body. (More about this momentarily.)

I also firmly believe a person should get as much nourishment as possible from food. Not pills, powders, or liquids concocted in laboratories. Real food. Food made by collaborative efforts of sunlight, water, and soil. At times supplements may be helpful or even necessary. But to me, prolonged reliance on manmade supplements signals an unhealthy diet, one that cannot lead to vigorous health over the long term.

Nutrition experts widely agree that food is the best source for human nourishment. The US government's Dietary Guidelines note that "nutrients should come primarily from foods."[4] The Mayo Clinic website points to three primary benefits of foods, compared with nutritional supplements. First, foods offer superior nutrition ("whole foods are complex, containing a variety of the micronutrients your body needs"). Second, foods contain necessary fiber.

Finally, real, minimally processed foods also hold various protective compounds ("whole foods contain other substances important for good health," including phytochemicals and antioxidants).[5] In *Nourishment*, Dr. Provenza even argues that these "secondary compounds" may be the most important parts of our diet for maintaining health and fighting off disease. Increasingly, the ancient wisdom of eating a diversity of whole, unprocessed foods is regarded as the key to good nutrition. Dr. Yudkin would have been pleased.

Despite the annual avalanche of books and articles telling us what to eat, nutrition science is in its infancy. More is unknown than known. To me, this gap in our understanding is the most compelling reason to aim for a diet rich in essential nutrients. We simply don't know whether dietary supplements can ever adequately replace the nutrients our forebears derived from foods. New York University nutrition professor Marion Nestle notes: "Clinical trials rarely show much benefit from taking [nutritional] supplements and . . . sometimes they show harm."[6]

I was fortunate to have gotten off to a solid start. My parents considered exercise and eating well the fundamentals for good health. My mother took great care in feeding us. She spent a lot of time with her garden, made frequent trips to local farms, cooked every dinner from scratch, baked bread, canned fruits, made yogurt. She taught her children how to grow, preserve, and prepare food. She also showed us, every day, the simple pleasure of sitting around the table with those you love, sharing conversation and a hearty, homecooked meal.

I strive to have my diet grounded on fresh and fermented foods—well raised, locally sourced, and simply prepared. Over the past decade, after passing the 40-year mark, I consciously increased eggs and dairy products to get more good fats and protein. Simultaneously, I reduced my sweets and flour. I spent a year from 2013 to 2014 on a "sugar fast," in which I abstained entirely from anything with added sweeteners. I eat more fruits, greens, and vegetables than ever. I include plenty of fats and fatty foods—including nuts, avocados, olive oil, coconut oil, and butter—and I consider them essential to good health.

I have always consumed bovine milk and fat in the form of liquid milk, yogurt, cheese, butter, and, of course, the occasional bowl of ice cream. But I refrained from eating meat for over three decades. This included 16 years while married to a cattleman and living and working on a ranch. Understandably, this often surprised people. But last year I changed that part of my diet, as well. When I turned 50, I began to re-evaluate my diet. I wanted to make sure I was doing everything possible to stay in optimal health. Bone loss worried me. Muscle loss worried me. Weight gain worried me. Those were all things I read about and saw all around me in women in their 50s. I was starting to experience them, too. My weight had increased, my muscles were shrinking. I had my bone density tested at Kaiser Permanente and was alarmed to learn I had osteopenia, the pre-cursor to osteoporosis.

As an environmental lawyer, rancher, and student of sustainability, I had come to view animals as an essential part of an environmentally optimal food system. As a researcher of nutrition, I had come to consider omnivory the ideal human diet. For years, I had been brushing away the quietly gnawing feeling my own vegetarian diet was less than optimal. I continued following a meatless diet, partly from habit. I didn't feel a physical urge to eat meat so it was easy to believe my body did not need it and just stay the course. If I'm perfectly frank I also didn't feel like mustering the energy to grapple with the implications of abandoning a diet I'd strictly followed for so many years.

But in August 2019, I took the plunge and had my first bite of meat in over 30 years. It was a burger, a real beef burger from cattle raised on our ranch, and cooked on the grill by my husband, Bill. As I raised it toward my mouth it smelled good and nourishing to me. I felt a twinge of trepidation. Was I going to regret this? But when I took that first bite it just felt good. The beef was delicious. I actually felt a wave of relief. I realized at that moment my diet was actually more in accord with my own values and beliefs when I ate meat than when I refrained from it. Eating that burger made me feel I was reclaiming control over my health as I moved into my 50s and beyond.

Obviously, as someone who lived more than three decades as a vegetarian, I respect a person's choice to abstain from meat. But if a person's rationale for not eating beef is based on a belief that it's necessarily bad for the environment or human health, it's probably poorly informed. Nothing about cattle raising is inherently harmful to the environment. Ecological injury from cattle is due to poor management. Similarly, fears that beef is bad for our health are proving unfounded. The benefits of eating meat, on the other hand, have been known since ancient times.

"Our primitive ancestors subsisted on a diet composed largely of meat and fat, augmented with vegetables, fruit, seeds and nuts," Sally Fallon writes in her book *Nourishing Traditions*. "Studies of their remains reveal that they had excellent bone structure, heavy musculature and flawless teeth." Fallon points to a wealth of anthropological and archeological research showing the meat- and fat-laden diets of early and traditional peoples contributed to their vigor. Early Mayan remains, from the time when meat was readily available, show adult male skeletons averaging eight centimeters taller than later remains, when meat became scarce. Russians in the Caucasus Mountains, who have the highest proportion of centenarians in the world, consume ample fatty meats and dairy. Populations in Soviet Georgian with the most meat and fat in their diets are those with the greatest longevity. "Equadorans from Vilcabamba, also known for their long lifespans, favor whole milk and fatty pork," Fallon explains.[7]

None of this is surprising, according to Fallon, because without animal products it is difficult to obtain adequate protein and minerals in a form the body can use. Animal-based foods not only supply protein and minerals, they provide the necessary fat-soluble catalysts needed for mineral absorption. "Zinc, iron, calcium and other minerals from animal sources are more easily and readily absorbed."[8]

Modern nutrition science, as imperfect as it is, has helped document many of beef's benefits. A 2013 article in *The Guardian* calls beef "one of the most nutritious foods" available, noting "beef has appetite-sating high-quality protein, which has all the essential amino acids needed (isoleucine, leucine, lysine, methionine, phenylalanine, threonine, tryptophan, valine and more) to build muscle and bone." Beef, it points out, is also a great source of B vitamins, iron, and zinc.[9] To that list I would add vitamin D, found in very few foods, yet available in the most bio-available form in beef liver.

As I'll discuss more a bit later, the most nutritious and wholesome beef you can eat is from cattle raised entirely on grass and other forages. Yet even standard grocery store variety beef is crammed full of nourishment. It takes just one 3-ounce piece of grain-finished beef to provide the following part of the US recommended daily allowance for an adult male:

Protein: 23 grams (45% of USRDA)
Minerals: 13% iron; 5% magnesium; 42% zinc; 4% copper;
    20% phosphorus; 24% selenium
B vitamins: 5% thiamin (B_1); 10% riboflavin (B_2); 16% niacin (B_3);
    14% pyridoxine (B_6); 54% cyanocobalimin (B_{12})

Beef organ meat provides an even more nutrient-dense meal. It truly merits the "superfood" label. Just 3 ounces offers the following:

Protein: 21 grams (43% USRDA)
Vitamin C: 71%
Minerals: 186% iron; 4% magnesium; 16% zinc; 39% copper;
    26% phosphorus; 7% potassium; 111% selenium
B vitamins: 3% thiamin (B_1); 15% riboflavin (B_2); 24% niacin (B_3);
    7% pantothenic acid (B_5); 2% pyridoxine (B_6); 71% cyano-
    cobalimin (B_{12})[10]

Anyone watching their weight should also know either of those pieces of beef also contains zero carbohydrates, and ranks zero on the glycemic index. These latter qualities make beef an exceptionally good choice for dieters, diabetics, and anyone over 40. Although I do not think eating naturally occurring saturated fat is a health concern, many people will also be interested to know the steak contains just 16 percent of the daily saturated fat allowance, and the organ meat contains just 6 percent. Beef truly is among the most nutritious foods available.

Nutrients in Beef, Uniquely Bio-Available

Now let's dig a little deeper into beef's benefits, starting with one of the most obvious—iron. As the nutritional content shows, beef is loaded with iron. A small steak has 13 percent of the recommended daily allowance, while organ meats have a whopping 186 percent. Even the steak is higher in iron than lamb, with more than double the content of turkey meat, and three to four times more than any other commonly available meats. The iron found in meat is especially valuable because it is highly usable by the human body.

Insufficient iron is a huge human health problem. "Iron deficiency is the most common and widespread nutritional disorder in the world," according to the World Health Organization (WHO). Up to 80 percent of the world's people are estimated to be iron-deficient, 30 percent afflicted by full-blown anemia. Women, especially pregnant ones, and children are hardest hit. In developing countries, half of pregnant woman and about 40 percent of preschool children are estimated to be severely iron-deficient. At the same time, says WHO, lack of iron is "the only nutrient deficiency which is also significantly prevalent in industrialized countries."[11]

In the United States, iron deficiency is widespread among girls and women, especially those of lower income and those who are pregnant. One study showed more than 80 percent of women 20 to 49 years of age were failing to get recommended amounts of iron from their diets.[12] This is especially concerning because most vegans (79 percent in the United States, for example) are female.

Anemia is severe iron deficiency. It's a condition in which the body has fewer-than-normal red blood cells in the blood, or the red blood cells have insufficient hemoglobin (the protein that carries oxygen from the lungs to the rest of the body). In the short term, iron deficiency causes low energy levels, and hinders memory and other mental functions.[13] Longer term, the medical consequences of anemia can be serious, especially for pregnant women. For all people afflicted, it can impair organ function. For pregnant women, it can harm physical and cognitive fetal development. Anemia also contributes to 20 percent of all maternal deaths.

The National Institutes of Health reports "anemia among lower income pregnant women has remained the same, at about 30 percent, since the 1980s."[14] More than three million Americans, mostly women and people with chronic diseases, have chronic anemia.[15]

As these figures make clear, women are the most at risk for iron deficiency. Women of child-bearing age need the most iron (the US recommended daily allowance is 18 milligrams for a woman between 18 and 50). Pregnant women need nearly twice that (around 30 mg).[16]

The dietary problem is not so much finding foods containing iron. It's the human body's capacity to *utilize* the iron. Iron intake is not equal to iron absorption, advises the Cleveland Clinic.[17] Iron comes in two distinct forms: heme (bound to hemoglobin) and nonheme (not bound to hemoglobin). Meat, especially red meats, contains large portions of heme iron, whereas plants with iron, such as spinach, lentils, and beans, contain only nonheme iron. Nonheme iron is also the type added to iron-enriched and iron-fortified foods. Most iron in modern diets is nonheme. However, "heme iron is absorbed better than nonheme iron," according to the National Institutes of Health.[18] Because the iron found in plants is poorly absorbed, NIH advises that vegans and vegetarians need *twice* as much dietary iron each day as meat eaters. "Absorption of iron into the body is greatest with meat sources of iron," Cleveland Clinic similarly advises.[19]

More specifically, NIH notes that the "absorption of heme iron from meat proteins is efficient" and is not much influenced by other foods a person eats. In contrast, NIH says, only "2 percent to 20 percent of nonheme iron in plant foods such as rice, maize, black beans, soybeans and wheat is absorbed," and its "absorption is significantly influenced by various food components." All of which makes it very difficult to get adequate iron exclusively from plants. Meat provides ample iron, and iron of better quality.

It's also noteworthy that foods help or hinder absorption of all types of iron. Meat not only provides heme iron but also assists the body in absorbing nonheme iron. On the other hand, absorption of nonheme iron is actually hampered by some vegetable protein sources, including soybeans.[20] "The reason for this is not well understood," notes Harold McGee about the differing absorption rates.[21] But clearly, a person who regularly eats red meat can more easily avoid iron deficiency than someone who does not.

The challenge of getting enough iron without red meat is real and personal to me. In both of my pregnancies, blood tests revealed I had dangerously low iron levels. In the first pregnancy, I was able, with a daily prenatal supplement and disciplined, carefully planned eating, to raise my iron levels to just barely within the acceptable range. In the second pregnancy, however, even with what seemed like Herculean efforts on my part, plant-based foods and prenatal vitamins could not raise my body's iron to a level that would be safe for my baby and me. Once I began adding a daily liquid iron supplement, it was just barely enough. It was very fortunate I paid attention and took the issue seriously. Just after my second son was born, I had post-delivery bleeding that necessitated emergency surgery. The hemorrhaging was severe enough that, doctors later informed me, had I not been at a modern medical facility I might well have died. The iron my body had stored in the preceding months was important in my survival and recovery. If I had been eating red meat, however, I probably would never have faced an iron shortage at all. My advice to vegetarian pregnant women is to pay close attention to this issue. Non-vegetarian pregnant women are well advised to regularly include beef in their meals.

An interesting side note about iron. Following the general pattern that what we evolved with works best: Iron in breast milk is far better absorbed by infants than iron in manufactured formula. "It is estimated that infants can use greater than 50% of the iron in breast milk as compared to less than 12% of the iron in infant formula," according to NIH.[22] "Reasons for the high bioavailability of iron in breast milk are unknown," states a hematology textbook.[23] (Just another reminder of how little we truly understand human nutrition, and yet another important reason to pay attention to nature's nutritional wisdom, especially the value of breastfeeding.)

As beef's nutritional information shows, just a 3-ounce serving of steak supplies nearly half the zinc needs of an adult. Even compared with other red meats, beef is particularly rich in zinc: It contains about a third more than is found in lamb, and more than twice as much as in veal or pork.

Zinc is essential to good health, according to the National Institutes of Health. The mineral is present in all cells throughout the body. It's important in immune system function, and protein and DNA synthesis. In pregnancy, infancy, and childhood alike—zinc is essential for development, according to NIH. Zinc aids in the healing of wounds and is important for functioning of the senses of taste and smell.[24]

Globally, zinc deficiency is widespread. The World Health Organization estimates about one-third of the world's population is short on zinc.[25] Some research suggests it is also common in the United States. One study showed 40 percent of US adolescent girls and adult women were failing to get the recommended daily allowance of zinc. Another showed 60 percent of men over the age of 70 failing to get adequate zinc.[26]

As with iron, the zinc found in beef is especially usable by the human body. According to the United Nation's Food and Agriculture Organization, studies show about twice as much zinc is absorbed from diets containing meat compared with those without meat. The reason for this is interesting. Plant foods with high levels of zinc (including legumes, whole grains, seeds, and nuts) also are high in phytate, a substance that inhibits mineral absorption. Notably, though, just as with iron, the availability of zinc from plant foods can be improved by including animal protein in the diet, FAO notes.[27]

Food for Our Bones, Muscles, and Brains

Protein is another good reason to eat beef. As noted earlier, a small portion of beef provides nearly half the recommended daily allowance of protein. In the developing world, protein deficiencies are widespread. Generally speaking, most Americans are not considered protein-deficient. Still, in all countries various situations warrant a diet richer in protein and fat and lower in carbohydrates. As discussed earlier, lowering carbohydrates helps control diabetes and assists in weight loss. Just with iron and zinc, the protein contained in beef is high quality, easily usable by the human body. Good-quality protein is always important for health, and even more so as we age.

This point was reinforced by a study conducted in Ohasama, Japan, and published in the March 2014 *Journal of the American Geriatrics Society*. It found that "after adjustment for putative confounding factors, men in the highest quartile of animal protein intake had significantly lower risk of higher-level functional decline than those in the lowest quartile." Notably, the research found no benefit to either gender from plant protein. The study authors conclude: "Higher quality protein, particularly animal protein, was associated with lower risk of decline in higher-level functional capacity in older men."[28]

Research published in 2018 also suggests that older people can avoid muscle loss by adding protein to their diets along with weight training. Researchers at McMaster University in Ontario reviewed 49 high-quality past experiments that had studied a total of 1,863 people, including men and women. The review found that eating more protein—well above currently recommended levels—can significantly augment the effects of lifting weights, especially for people past the age of 40. "Men and women who ate more protein while weight training did develop larger, stronger muscles than those who did not." Those who ate more protein "gained an extra 10 percent or so in strength and about 25 percent in muscle mass compared to the control groups." The amount recommended by the study was about 1.6 grams of protein per kilogram of body weight per day, or about 130 grams of protein a day for a 175-pound man. Notably, the number is quite a bit higher than current federal dietary recommendations, which suggest about 56 grams of protein a day for men and 46 grams a day for women.[29]

Meat's high-quality protein is critical to preventing loss of both muscle and bone. After about age 40, our bodies have a harder time maintaining muscle mass, and our bones tend to become weaker. The two problems are closely connected. A 2008 study in the *American Journal of Clinical Nutrition* notes, "Maintenance of adequate bone strength and density with aging is highly dependent on the maintenance of adequate muscle mass and function, which is in turn dependent on adequate intake of high-quality protein."[30] The researchers conclude that despite some claims to the contrary, "higher protein diets are actually associated with greater bone mass and fewer fractures when calcium intake is adequate."

The study's authors urged retraining for medical professionals about the importance of protein in diets of older patients.

Similar results were obtained by researchers at Hebrew SeniorLife's Institute for Aging Research (IFAR), affiliated with Harvard Medical School. In 2000, it published a study in the *Journal of Bone and Mineral Research* concluding that protein intake generally, and animal protein in particular, promotes bone health. The researchers found lower protein consumption significantly related to bone loss, "suggesting that protein intake is important in maintaining bone or minimizing bone loss in elderly people." An IFAR article on the study notes that the research conformed with two earlier randomized controlled trials showing that increased protein intake dramatically improved outcomes after hip fractures.[31]

Osteoporosis is a particularly great danger for women, explains Miriam Nelson, PhD, author of *Strong Women, Strong Bones*. "Starting around age 35 we lose up to 1 percent of our bone mass each year," her book notes. "These losses accelerate rapidly after menopause."[32] For reasons not fully understood, Nelson says, muscle mass also affects bone density. The more muscle mass we have, the less bone density we lose. And she advises, "protein is important for bone health." Nelson warns, "About 30 percent of women don't get enough protein, which adds to their risk for osteoporosis."[33]

A less well-known factor in bone thinning is widespread vitamin K_2 deficiency. In her book *Vitamin K_2 and the Calcium Paradox*, nutritionist Kate Rheaume-Bleue makes a compelling case that calcium is only one component necessary for bone health. Equally important, she urges, are fat-soluble vitamins A, D, and K_2. "The true test of a nutritious diet is that it provides these vitamins," Rheaume-Bleue writes. K_2, only recently fully recognized by modern science, is found almost exclusively in animal-based foods. Milk, meat, and eggs from animals living on grass are the best source, she advises. She blames current epidemic osteoporosis levels largely on the modern diet's shortage of vitamin K_2. "[T]he gradual industrialization of our food supply that removed animals from pasture siphoned off menaquinone (K_2) in the process," she explains. "This was further complicated by the introduction of trans fats and the decades-long crusade against the kinds of foods that are highest in vitamin K_2—egg yolks, cheese, and butter."[34] Here again, the core of our modern health problems connects to abandoning real food and replacing it with things that have been industrially produced.

For people of all ages, beef provides nutrients valuable to optimal brain function. As a 2008 article about "brain food" in the highly respected journal *Nature* makes clear, beef contains important nutrients involved in brain function: iron, copper, selenium, zinc, B_6 and B_{12} vitamins, choline, vitamin D, and vitamin C.[35] *Nourishing Traditions* argues that infants and children need cholesterol, like that contained in bovine fat, for proper brain development.[36] A 2020 research paper from University of Cincinnati physicians states: "As with the fetus, neonatal mammalian cells also require significant cholesterol for normal cellular function. Infants fed human milk receive much greater quantities of cholesterol than those fed commercial formulas." The study notes human milk contains 10 to 15 mg/dl of cholesterol, providing an average daily cholesterol intake of around 75 mg per day for a breastfed 4 kg new-born. Cow-milk-based formulas contain much less, giving an average daily cholesterol intake of approximately 9 mg per day. Soy-milk-based formulas contain no cholesterol. Young babies obtain cholesterol from their own synthesis as well as dietary cholesterol, the study notes. Breast milk is "the largest contributor of exogenous cholesterol."[37] The long-term consequences of depriving fetuses and infants of cholesterol are unknown and, clearly, a cause for concern.

Cattle raised entirely on grass also have vitamin E, omega-3 fatty acids, and beta-carotene in their flesh, fat, and milk.[38]

Best Food Source for Vitamin D

In recent years, heightened attention has been paid to vitamin D. "In the United States and Europe, it is estimated that more than two-thirds of the population is deficient in vitamin D," *The New York Times* reported in 2014.[39] Increasingly, vitamin D shortfalls are being connected to a wide variety of health problems—everything from diabetes, stroke, hypertension, and autoimmune disorders to heart disease.[40]

Vitamin D is found in few foods. The most usable form comes from the sun. For most of the year, in most latitudes where humans reside, you can get adequate vitamin D by spending some time outside every few days. Today's epidemic of vitamin D shortages is yet another side effect of modern lifestyles spent indoors in front of computers and TV screens. It is exacerbated by the overly zealous application of sunscreens intended to decrease skin damage from exposure to the sun's UV rays. But it becomes harder to get adequate vitamin D as we get older.[41] "If a 65-year-old woman and her 35-year-old daughter take a 10-minute walk on a sunny day, the mother's skin will manufacture only a third of the amount of vitamin D that her daughter's does," according to Dr. Nelson.

For those who cannot get enough sunshine, some of the best foods available to help shore up vitamin D levels are products of cattle. One is whole milk (especially from cows kept on grass), which is usually fortified with vitamin D_3 (typically 100 IU in an 8-ounce glass). Another is beef liver, which can provide 19 IU per 100 grams. This is only a small portion of a person's needs, but it could be a vital part of a strategy to improve vitamin D levels. Few people realize red meat helps with vitamin D in another way. According to nutrition expert Chris Kresser, red meat "contains a vitamin D metabolite called 25-hydroxycholecalciferol, which is assimilated much more quickly and easily than other dietary forms of vitamin D."[42] There is even evidence that meat provides protection against rickets, a disease caused by severe vitamin D deficiency. This reinforces the idea the vitamin D in meat is "uniquely absorbable and useful to the human body."[43]

B as in Beef

Beef is also an excellent source of B vitamins. If you're unsure what's meant by "B vitamins," you are not alone. The answer is complicated and has changed quite a bit over the years. Once regarded as a single vitamin, scientists now consider the B's a complex of 10 or 11 (depending on whether you include choline) chemically distinct vitamins often found in the same foods and frequently functioning together as a group.[44] The B vitamins are now considered vital, affecting the functioning of "every organ system and all aspects of our health."[45] Specifically, they are thought to be important in methylation, the process by which protein and DNA are produced and sustained in the body; important in heart health; and essential in creating neurotransmitters responsible for brain function and mood.[46]

Deficiencies of the B vitamins were once considered rare. Yet this view is shifting. A school of thought lately emerging is that deficiencies may be widespread—affecting as much as 60 percent of the US population.[47]

Beef and beef organs are good sources of nearly all B vitamins, including B_1 (thiamin), B_2 (riboflavin), and B_5 (pantothenic acid), as well as B_{10} (PABA), inositol, and choline. The National Academy of Sciences recommends a daily choline intake of 550 mg for men and 425 mg for women. Just 3 ounces of ground beef contains 67.4 mg of choline (12 and 16 percent, respectively, of recommended amounts).[48] B_6 deficiency is believed to contribute to serious health problems, including cardiovascular disease and cognitive decline. But B_6 is found in both beef and beef organ meats.[49]

Vitamin B_{12} (cobalamine), a substance occurring only in animal-based foods, is of particular concern. B_{12} deficiency affects between 1.5 and 15 percent of Americans. In certain segments of the population, deficiency rates are much higher. Among those most at risk are people over age 50, people with pernicious anemia, and people with gastrointestinal disorders (including Crohn's disease and celiac disease)—all of whom may have difficulty absorbing B_{12}.

Vegans are at particularly high risk. A 2010 Oxford University study found that over half of vegans and 7 percent of vegetarians are B_{12}-deficient.[50] Similarly, a 2003 study published in the *American Journal of Clinical Nutrition* tracked 174 apparently healthy people living in Germany and the Netherlands. The study found 92 percent of the vegans studied had vitamin B_{12} deficiency; two-thirds of vegetarians were deficient. Meanwhile, only 5 percent of those who consumed meats had vitamin B_{12} deficiency.[51]

The stakes are particularly high for pregnant and nursing women, and their babies. "When pregnant women and women who breast-feed their babies are strict vegetarians or vegans, their babies might also not get enough vitamin B_{12}," NIH advises. Infants who get insufficient vitamin B_{12} may experience developmental problems and a debilitating condition called megaloblastic anemia (a disorder characterized by large but immature and incompletely developed red blood cells that do not function like healthy cells).

In adults, B_{12} deficiency can cause tiredness, weakness, depression, dementia, and megaloblastic anemia. Up to 30 percent of patients hospitalized for depression are B_{12}-deficient.[52] Long term, the deficiency can damage the nervous system. Among otherwise healthy people who regularly eat meat, B_{12} deficiencies are uncommon. All beef contains B_{12}, and beef liver is one of the best possible sources of the nutrient, according to NIH.[53]

By now you have undoubtedly noticed a pattern: Beef organs stand out as jam-packed with nutrients in a highly usable form. They are simply one of the best sources of some of the most vital nutrients in the human diet. Integrative health practitioner Chris Kresser, who calls organ meats "nature's most potent super-foods," points out in some traditional cultures, *only* the organ meats were consumed while the lean muscle meats we mostly eat in the United States today were discarded or given to the dogs.[54] As a general rule, Kresser says, "organ meats are between 10 and 100 times higher in nutrients than corresponding muscle meats." He also notes while people often worry that environmental toxins might build up in cattle livers, research does not bear this out. Well-raised, grass-based animals, which are not fed concentrates, are especially unlikely to have toxins in their flesh or organs. Kresser advises his patients to "eat meat and organ meats from animals that have been raised on fresh pasture without hormones, antibiotics or commercial feed. Pasture-raised animal products are much higher in nutrients . . ."

Comparing Grass-Fed with Grain-Fed

While this is certainly true for pasture-raised eggs, when it comes to beef and dairy products, Kresser's comment may be somewhat overstated. (More specifics on that point in a moment.) The reasons to seek out dairy and meat from cattle raised on grass are compelling, nonetheless. The majority of cattle raised for beef in this country are implanted with growth hormones (usually both on the ranch and again at the feedlot), and most are also fed antibiotics and other pharmaceuticals as part of their feedlot ration. But I have never heard of a farm or ranch raising truly grass-fed cattle that engages in either practice. To me, when combined with the environmental problems associated with large feedlots discussed in the first half of this book, and animal welfare considerations, the reasons for seeking foods from cattle raised wholly on grass are compelling.

For a host of reasons, I strongly favor grass-based farming and ranching. It's more ecological. It creates a more humane, healthy living environment for animals. It's a better work environment for the humans tending the animals. It generates healthier, safer food.

Life on the pasture provides manifold benefits even for omnivorous animals like turkeys, chickens, and pigs.[55] I have long argued against continual indoor confinement of farm animals: All creatures need daily exercise, fresh air, and sunshine to be healthy and have a decent life. Animals must have decent lives if they are to provide flavorful, nourishing food. Humane conditions for animals mean better work environments for the humans taking care of them, as well.

The way an animal is raised also directly affects the quality of the food produced. "Rapid, confined growth favors the production of white muscle fibers, so modern meats are relatively pale," writes famed food scientist Harold McGee.[56] "They're also tender, because the animal gets little exercise, because rapid growth means that their connective tissue collagen is continuously taken apart and rebuilt . . ." On the other hand, "Full-flavored meat comes from animals that have led a full life," McGee notes. "[W]ell-exercised muscle with a high proportion of red fibers (chicken leg, beef) makes more flavorful meat than less exercised predominantly white-fibered muscle (chicken breast, veal) . . . This connection between exercise and flavor has been known for a very long time."[57]

McGee also notes the connection between fat and flavor. The modern tendency to breed for leanness and slaughter animals younger and lighter has diminished eating quality. "Thanks to the industrial drive toward greater efficiency," he writes, "and consumer worries about animal fats, meat has been getting younger and leaner, and therefore more prone to end up dry and flavorless."[58]

Feeding antibiotics or other drugs to cattle, as well as using growth hormones, causes environmental contamination and raises the specter of food tainted with drugs, hormones, and antibiotic-resistant bacteria. There is evidence that keeping cattle on forages (rather than concentrated feeds like grain) up to the point of slaughter can make beef safer, by reducing the occurrence of the dangerous *E. coli* strain 0157.H7.[59]

Nonetheless, I want to take a moment to correct a couple of modern myths surrounding cattle feeding. Contrary to what's often suggested in the popular press, there's nothing inherently wrong with feeding cattle a certain amount of grain. As discussed earlier, modern cattle breeds—for both meat and milk—are descended from a long line of domesticated cattle fed grain at least some of the time. Cattle sufficiently mature and given grain in appropriate amounts are well equipped to tolerate it.

Keeping all of that in mind, I consider grass the optimal diet and environment for cattle. It provides animals by far the most natural life and creates the safest, highest-eating-quality, and most nutritious food. The mainstream beef industry has funded several health and environmental studies purporting to show there's no difference between typical feedlot beef and grass-fed beef. These reports, designed to justify status quo methods, lack credibility. The science showing superior nutritional profiles from grass-fed animals is sound. Numerically, the differences in nutrient content are not large. Yet there is a consistent, across-the-board nutritional advantage to beef and milk from cattle raised entirely on grass and other forages.

To date, the most comprehensive review of studies comparing nutrition of foods from grass-fed versus grain-fed cattle was done by Union of Concerned Scientists' senior scientist Kate Clancy, who holds a doctorate in nutrition science. Regrettably, the report, titled *Greener Pastures*, is largely focused on fats, which, clearly, I do not consider a health concern. Nonetheless, the report is credible and important. Overall, it concludes: "Meat from pasture-raised cattle contains less total fat than meat from conventionally raised animals, and both meat and milk from pasture-raised animals contain higher levels of certain fats that appear to provide health benefits." The nutritional differences, the report notes, result from "the chemical differences between forage and grains, and the complex ways in which ruminant animals such as cattle process these feeds."[60]

More specifically, the report found that all the grass-fed steak in the studies reviewed could be labeled "lean" or "extra lean"; some could also be labeled "low-fat" (no more than 3 grams of total fat per serving).[61] Steak and ground beef from grass-fed cattle are almost always lower in total fat; steak from grass-fed cattle tends to have higher levels of the omega-3 fatty acid alpha-linolenic acid (ALA); and ground beef from grass-fed cattle usually has higher levels of conjugated linoleic acid (CLA). Meanwhile, milk from pasture-raised cattle tends to have higher levels of ALA and has consistently higher levels of CLA. ALA is believed to reduce heart disease risk. CLA, as mentioned earlier, has positive effects on cancer, as well as heart disease and the overall immune system. "CLA has been associated in animal and laboratory studies with an impressive array of health benefits . . . [relating to] atherosclerosis, diabetes, immune function, and body composition."[62]

Greener Pastures also found strong evidence of higher vitamin content in foods from grass-fed cattle. "[A] substantial number of studies that measured alpha-tocopherol showed that meat and milk from pasture-raised animals had levels of vitamin E significantly higher than meat and milk from animals fed grain," the report concludes.[63] It found the same pattern for beta-carotene, the plant precursor of vitamin A. "Forages contain substantial amounts of beta-carotene and grains contain very little," the report explains. "[B]ecause beta-carotene is easily oxidized, stored forages have a significantly lower concentration of the substance."[64] These factors make the forage-based diet healthier for cattle and result in higher levels of both vitamin E and beta-carotene in foods. However, it's important to acknowledge that both nutrients are present only in small amounts in beef and milk, even from animals raised on grass.

In the years since the *Greener Pastures* report was published, a wealth of new research has considered the possible benefits of raising cattle entirely on grass. Most has reached conclusions remarkably in keeping with that report. Research continues to show that grain feeding has a negative effect on the presence of omega-3s in cattle fat. This is because the fatty acids found in grass are mostly—60 percent—omega-3s.[65] When cattle are raised entirely on grass, their fat is about 3 percent omega-3 fat. However, the omega-3 level steadily drops the longer the animal is in a feedlot. By day 196, the level reaches zero.[66]

It is not surprising, then, that several studies have shown notable differences in the fats found in grass-fed versus grain-fed beef. Research done by Clemson University and the Department of Agriculture in 2009 found that compared with grain-finished beef, grass-fed beef is higher in CLA and total omega-3 fatty acids; it also has a far better ratio of omega-6 to omega-3 fatty acids (1.65 for grass-fed, compared with 4.84 for grain-fed beef). In addition, the researchers found that grass-fed beef has higher levels of each of the following nutrients: beta-carotene, vitamin E (alpha-tocopherol), calcium, magnesium, potassium, and the B vitamins thiamin and riboflavin.[67]

A 2010 study done collaboratively by California State University and the University of California and published in *Nutrition Journal* looked closely at the ratio of omega-6 to omega-3 fats.[68] It notes that while a healthy diet should consist of roughly one to four times more omega-6 fatty acids than omega-3 fatty acids, "[t]he typical American diet tends to contain 11 to 30 times more omega-6 fatty acids than omega-3s." The report presents evidence of significant differences in omega-6:omega-3 ratios between grass-fed and grain-fed beef. Considering all studies reviewed, it found an overall average ratio of 1.53 for grass-fed beef and 7.65 for grain-fed. Thus, the omega-6:omega-3 ratio is far more favorable for grass-fed beef.

A 2011 study in the *British Journal of Nutrition* showed that even moderate consumption of grass-fed meats for just four weeks benefited the omega-6:omega-3 ratio of the human subjects.[69] In 2018, University of Minnesota researchers published a three-year study in which they quantified the fatty acid profile in milk of over 1,000 samples of milk from cows fed a 100 percent forage-based diet, comparing it with profiles of milk from cows in conventional dairies. Researchers found grass-fed milk provided by far the highest level of omega-3s: 0.05 gram per 100 grams of milk, compared with 0.02 g/100 g in conventional milk, a 147 percent increase in omega-3s. Grass-fed milk also had 52 percent less omega-6 than conventional milk.[70]

Research from Ireland in 2019 concluded that dairy cow diets, pasture versus concentrates, "have a significant effect on the composition and quality of milk." It found milk from pasture-based cows to have higher fat and protein content with improved nutritional status (including a better n-6:n-3 ratio). It also found that pasture feeding results in milk with higher concentrations of a variety of beneficial nutrients including vaccenic acid, CLA, beta-carotene, and alpha-linolenic acid.[71]

The 2010 California study (UC and Cal State collaboration) also considered a newly identified nutrient that may play an important role in keeping a human healthy. Glutathione (GT) is "a relatively new protein identified in foods" that "functions as an antioxidant." GT works within cells, quenching free radicals and protecting cells from oxidized lipids or proteins and preventing damage to DNA. Fresh vegetables and beef are both high in GT, the report notes, and "[b]ecause GT compounds are elevated in lush green forages, grass-fed beef is particularly high in GT as compared to grain-fed . . . Grass-only diets improve the oxidative enzyme concentration in beef, protecting the muscle lipids against oxidation as well as providing the beef consumer with an additional source of antioxidant compounds."[72] From a variety of perspectives, evidence is mounting that grass-fed beef and dairy products are healthier and more nourishing than grain-fed. Since first writing the first edition of this book, grass-fed beef, milk, and cheese (and eggs) have all become more widely available and less expensive. While I consider all American beef and milk a good dietary choice, I believe we will continue to see more evidence in coming years that grass-fed beef and dairy are best.

Really Tasty Food

When all the food, health, and environment arguments are said and done, whenever we consider what to eat we must come back to flavor. At least that's true for both my husband, Bill, and me. Gathering around the table for dinner with our boys is our most meaningful daily ritual. It's where we nourish ourselves while engaging in our most important conversations. We cherish meals shared with friends, as we do the occasional special dinner at Chez Panisse, or the restaurant of our dear friend Chef Duilio Valenti. Eating really good food—especially with people we love—is among life's sweetest pleasures. No matter how healthy grass-fed cheese or beef is, why eat it if it doesn't taste great? For us, this is a basic rule for every food we bring into our home. It's the main reason we decided to stop buying out-of-season strawberries and tomatoes.

When Bill first considered returning to raising cattle entirely on grass (as he'd done years earlier), he took small, tentative steps toward the venture. He wasn't going to do it at all if he could not find a way to consistently make beef that, as he says, "eats great, every time."

Bill had feasted on mouthwateringly delicious grass-fed beef while traveling in Argentina, Australia, South Africa, and New Zealand. It was rich, flavorful, and had just the right amount of chew. But much of the grass-fed beef he'd sampled in the United States simply did not stack up. It was often dry, tangy, and tough. Once at a white-tablecloth restaurant in the Midwest, the chef personally came to our table to proudly serve Bill grass-fed beef from the Scottish Highland cattle of a local farmer.

When we were alone Bill leaned over to me and whispered, "This is inedible." He ate it, of course. But only out of politeness and respect for the chef and the farmer. That kind of beef was what was worrying Bill. In recent years, though, he's been impressed with grass-fed beef in several regions of the United States.

Famed *New York Times* food writer Marian Burros has had a similar experience. "When I wrote about grass-fed beef in 2002 there were about 50 producers, and most of what they raised was not very good," she has written. By 2006, however, she found the grass-fed beef sector to have "taken giant steps." Burros could locate about 1,000 farmers and ranchers raising grass-fed beef, and, from an eating-quality vantage point, she felt "more of them are learning to get it right." Her verdict: "My tasting showed that with 100 percent grass-fed beef you can have it all: sustainable, more nutritious beef with clean, juicy, beefy flavor. (Because the beef has less fat, though, it must be cooked at lower temperatures and for less time.)"[73]

The idea that grass-fed beef, especially the way it is generally raised in the United States today, must be approached differently in the kitchen probably has some validity. Food writer Lynne Curry's book *Pure Beef* is based on this premise. Curry takes issue with the standard cooking advice for grass-fed beef (which she sums up as follows: "Don't cook it over high heat!" "Don't cook it past medium rare!" and "Don't salt the meat before cooking, or it will be dry!"). Nonetheless, she concludes that beef raised entirely on grass requires some special care. Her advice is this: Season it well; don't hesitate to brown it whenever you'd like; use an appropriate cooking method; and cook it a little shy of the serving temperature you want.[74] Curry does acknowledge, however, that grass-fed beef varies. Some ranchers have divined "the magic formula for producing pasture-raised beef that is out of this world." Lynne does not mention anyone by name, but I like to think she (who has been to our ranch and eaten our beef) includes the beef from our ranch in that elite group.

Here, for which I apologize in advance, I must brag a little. My husband, Bill, is as hardworking a person as you will ever meet. His passion for his work runs marrow-deep. Nothing is ever good enough unless it's perfect. From the perspective of his spouse, I can tell you this has both advantages and disadvantages. But the people who always benefit from Bill's relentless passion for excellence are his customers.

After having tasted grass-fed beef from around the United States and in various distant countries, Bill had a realization: When it was bad, grass-fed beef was very bad indeed; but when it was good, it was the best of all beef. In the late 1990s, he began giving serious consideration to raising his own totally grass-fed cattle. When I first met him in 2000, Bill was entertaining the notion of developing a line of 100 percent grass-fed beef for Niman Ranch, the company of which he was founder and CEO. He was being egged on somewhat by things he was reading and by conversations he was having with environmental advocates, writers, and chefs.

Foremost among them was Alice Waters, a world-famous restaurateur and longtime customer of Bill's. She had become convinced all beef should be raised entirely on grass. Yet like Bill, Waters was finding the eating quality highly variable. Based on her years working with Bill, she believed he could produce excellent grass-fed beef if he set his mind to it. As Bill and I became acquainted, I encouraged him in this pursuit, as well. I had read and seen enough to know raising cattle entirely on grass had myriad environmental advantages.

Aside from the logistics of it, Bill's only real hesitation (occasionally reinforced) was his concern about eating quality. He was determined to get involved with grass-fed beef only if he could *consistently* generate beef that tasted at least as good as beef from cattle finished on grain. This was the genesis of a quest that lasted several years. Before getting started, he'd queried longtime ranchers about what they thought it would take to raise cattle for great beef entirely on grass. And he'd spoken with chefs about their expectations and experiences on the meat side.

Over the next seven years, Bill raised, every year, a handful of cattle entirely on our Bolinas home ranch and exclusively on grass. Each year, he slaughtered them progressively, at different moments, according to the condition of the grasses and the body conditions of the animals. He took the cattle in very small groups (two or three at a time) when he considered each individual animal at its peak. Taking careful note of the animal's condition, he observed how it corresponded to differences in the fat and meat. As always, he personally transported each animal to the slaughterhouse and personally oversaw the entire kill, and the butchering.

Together with his colleagues at Niman Ranch, he would then methodically taste a variety of cuts from each of the carcasses. Additionally, each year, he organized tastings of the meat by chefs whose palates he particularly trusted.

His interviews and experiences over those years led to the development of a method. To some degree, we consider this method unique. At the same time, though, we recognize what we are doing is really just following age-old practices of good animal husbandry and ancient wisdom about when animals should be slaughtered. We are happy to share what we have learned with anyone who asks.

Making Great Grass-Fed Beef

Here are the basics of our method. Starting with the right breeds and lineages is essential. We have a herd of mother cows and bulls, all of British breeds (called *native cattle* in the trade). Every few years we have some Hereford bulls, but our mother cows (and most of our bulls) are predominantly Aberdeen-Angus. After having sampled meat from literally thousands of carcasses over several decades while at Niman Ranch, Bill is firmly convinced British cattle breeds consistently provide meat of the highest eating quality. As he says, "The British have always been a beef-loving culture. They focused on creating breeds that produced great-eating beef on grass." On top of that, Angus cattle are easy to work with: They are even-tempered, make excellent mothers, have few health problems, and have no horns.

Our cattle do their coupling in the winter, from mid-December to mid-February. They mate naturally and always seem to enjoy themselves immensely in the process. (If you think I am kidding, you should watch a bull nuzzling and licking a cow's neck in the breeding season.) In a typical year, over 95 percent of our cows become pregnant. Our goal is always to include in our herd only mother cows that need no assistance in birthing and only bulls that sire calves sized appropriately to make that possible.

A calf's gestation period is nine and a half months, so the calves are born the following fall—from September to December. We deliberately time the calving this way to be in harmony with our grasses. Mother cows need more nutrition as their young ones grow, and around here grasses are strong starting in late winter and into the spring. (Conversely, we have found when there is too much feed available for the mother cows at birthing time, the calves are more likely to get scours and the cows are more likely to have udder problems.)

Calves live with their mothers and the rest of the herd of mother cows for seven to nine months. (It truly takes a cow village to raise a calf. Mother cows collectively watch over the young, especially when it comes to protecting them from predators.) Calves are then separated from their mothers using a "fence line weaning" technique. This keeps the pairs in the same location separated only by a wooden fence, minimizing stress for both cow and calf. From that point forward, the weanlings live together as a group and are always kept on the best pastures due to the high nutritional demands of their growing bodies.

These offspring (other than those heifers we decide to keep, who eventually join the cow herd) will continue to live on pastures and rangelands as yearlings and two-year-olds. More precisely, our cattle will go to slaughter averaging between 24 and 30 months of age and, for steers, weighing between 1,250 and 1,350 pounds; for heifers, between 1,150 and 1,250 pounds.

Anyone who raises cattle will recognize those weights as indicating fully grown, fat cattle. We refer to them as "grass-fattened," and they truly are. Bill often compares the way cattle bodies deliberately fatten themselves, when the opportunity presents itself in abundant feed, to get through the lean times to the way a bear gorges itself on salmon to get through a winter in hibernation.

We consider these two things—maturity and fat—vital to creating the best beef. Just as the oft-repeated farming saying goes, "Old hens make the best soup," in all species age adds flavor. "According to a standard French [culinary] handbook, *Technologie Culinaire* (1995), the meat of [a beef] animal less than two years old is 'completely insipid' while meat 'at the summit of quality' comes from a steer three or four years of age," Harold McGee points out.[75] (Although I should note that, with a similar idea in mind, for the first two years, Bill slaughtered only animals that were at least three years old, and even some at four years. He found no improvement in the flavor compared with two-year-olds.) As you can discern from the earlier discussion, we embrace the notion that beef fat—especially from truly grass-raised cattle—is a healthful and delicious part of a human diet.

Every geography has a season for beef, just as it does for peaches, strawberries, and tomatoes. In California, with our Mediterranean climate, that time is just after the grasses begin to dry out—usually in May to June. In other parts of the United States, it will be shortly after the first hard frost—typically in October. As I wrote about in some detail in *Righteous Porkchop*, feedlots, where cattle are fed corn and other grains, were invented largely for the purpose of smoothing out the swells and dips in supply—eliminating beef's seasonality. Farmers, chefs, and other eaters of grass-fed beef should keep this principle in mind. Anyone who wants to eat great-tasting grass-fed beef (and beef at its nutritional peak) must seek beef from cattle slaughtered in the right season. You can eat beef fresh in that season and frozen (or otherwise preserved) for the rest of the year.

To state it briefly, then, we believe consistently great grass-fed beef requires the following elements: British-breed cattle, kept always on grass, raised to maturity (minimum 24 months), slaughtered just after the grass season has peaked and the cattle are in prime condition (read: fat).

Our cattle, both those raised for meat and the breeding animals, spend every day of their lives on pasture or natural rangelands. The sole exceptions to this are when we have an individual sick animal (who may be kept in a "sick pen" in the corrals until recovered), and during the weaning when, as I just mentioned, the calves will be kept in the corrals for several days so they can share a wooden fence line with their mothers, who stay on pasture. In a really good grass year, we might reverse that— the cows kept in the corrals and the calves on the other side of the fence grazing highly nutritious late-spring feed.

As for the feeding, the regimen is straightforward. We never feed any grain, nor any kind of by-products, nor drugs of any kind. (And it goes without saying, we never use any growth hormones.) From birth until weaning, calves nurse on their mothers and graze on pasture. Other than their mother's milk, the only thing our cattle ever eat is grass and other naturally occurring vegetation (like clover and vetch), along with a small amount of alfalfa hay, which is fed in pasture.

The hay we use accounts for less than 1 percent of their annual diet. We'd rather not feed hay at all because it needs to be grown elsewhere and transported here, which adds to the resources expended, and adds considerably to our input costs. However, the protein (we buy only hay deemed high-quality but that didn't test high enough for dairies— around 20 percent protein) is a helpful supplement to the mother cows' diets in the dry season, when the nutritional content of our forages is at its lowest. We could operate without hay. But it would mean less vigorous animals, perhaps more susceptibility to disease, and most likely reduced fertility rates. This reduction in herd health and fertility also translates to a greater environmental impact.[76] Regarding alfalfa feeding, the Pimentels note: "In contrast to most other crops, alfalfa needs little or no nitrogen fertilizer; like legumes, it is associated with nitrogen-fixing bacteria. Because nitrogen fertilizer is an energy-costly input, this savings helps keep alfalfa production relatively energy efficient."[77] We believe, on balance, this limited feeding of hay is the right thing to do.

All of the cattle going into the beef we sell under our own name (Bill Niman Farms) come from our own herd or from the herds of a few other ranches we know well and work closely with, and that follow the same feeding and husbandry practices. We never buy any cattle from sale yards or any other source other than ranchers we know personally. This approach assures us and our customers that the origins and history of all of our beef is fully known and transparent.

A lot of chefs have told us our beef is the best beef they've ever tasted. Not the best grass-fed beef, the best beef period. Internationally renowned chef Dan Barber called eating our beef "a transformative experience." We are always looking for ways to improve what we do but I think we've found a successful formula. We are proud of the way we raise our animals and the beef we produce. We know it is exceptionally nutritious, healthful, and safe. Equally important, it's delicious.

Beef Lasts

Finally, I'd like to mention an additional quality that makes beef a uniquely valuable human food. It's an important point that's often over-looked: Beef is much less perishable than other foods. Until the advent of refrigeration (in 1834), people around the world relied entirely on techniques like drying, smoking, salting, and pickling to preserve meats and other foods. Some methods enabled meat to be preserved for years.[78] Beef has always been one of the foods that stay good the longest.

Somewhat ironically, the very thing for which beef has been most often maligned of late—saturated fats—is much of the reason for its good keeping quality. Harold McGee explains: "Unsaturated fats are most susceptible to rancidity, which means that fish, poultry and game birds go bad most quickly. Beef has the most saturated fat and stable of all meat fats, and keeps the longest."[79]

McGee also explains, "the intact muscles of healthy livestock are generally free of microbes," and "the bacteria and molds that spoil meat are introduced during processing, usually from the animal's hide or the packaging-plant machinery." In contrast with beef, he notes, "poultry and fish are especially prone to spoilage because they are sold with their skin intact, and many bacteria persist despite washing."

These days, of course, freezing is the primary method of beef preservation, at least in the industrialized world. Butcher and writer Adam Danforth points out in his book *Butchering Beef* that beef is unique in both the quantity of meat provided and the length of time the meat stays good. He notes that the average yield from a whole carcass of beef is more than 400 pounds of boneless meat, making a single animal the source of a year's worth of beef for two average-sized families. "Furthermore, frozen beef stays palatable for a longer period of time than almost any other kind of meat, allowing you to extend your harvest well beyond one year."[80] We should choose beef well and wisely. Then dig in and enjoy.

What's the Matter with Beef?

GIVEN THIS BOOK'S TITLE, IT MAY SURPRISE SOME PEOPLE THAT I find quite a lot wrong with the ways most cattle in the United States today are raised and turned into meat. Most of these concerns were detailed in the pages in my previous book, *Righteous Porkchop*, and I will not restate them all here. A veritable flood of books and articles has been devoted to describing the "dark sides" of cattle ranching and beef. As I hope I've already made abundantly clear, I find much of what's been written to range from the poorly informed to the flat-out absurd. Yet some of the criticisms have validity. I want to briefly take note of those now.

Stated succinctly, problems fall into the following categories: the way cattle are managed on the land; substances they are fed; hormones and other drugs used to stimulate growth; polluting practices; wasted resources; long-distance transport of live animals; and animal handling at slaughter. Said another way, these are problems of land management, wasted resources, pollution, animal welfare, and food safety.

Before I became directly involved with cattle ranching—while working as an environmental lawyer—I had some distance from these concerns, much as I would in dealing with the pollution of any sector. Admittedly, at that time I viewed these issues more objectively. But my depth of understanding was shallower. Now, as a member of the community of people who raise cattle and sell beef, I have a far better grasp of the issues and feel more urgency about solving them. Each day that problematic practices continue in the cattle and beef industry, it gives a black eye to everyone who raises cattle or is involved in the production, sale, or preparation of beef.

Every self-help program under the sun says the first step to correcting your shortcomings is self-examination and acknowledgment. I regularly peruse several meat industry periodicals. In nearly every issue, I am struck by the current mode of dealing with consumer concerns—rarely acknowledging and often denying. Rather than fairly assessing a criticism then rallying everyone to fix it, again and again the meat industry falls back into a circle-the-wagons defensive posture. This has been true for beta-agonists, hormones, feeding of antibiotics, and many more issues. Too often, there's tone deafness. The beef industry frequently dismisses critics as people who simply "do not understand agriculture," or "don't care about feeding the world," or "need to be educated." More often than not, it demonstrates a near-total unwillingness to adapt or change.

Earlier, I cited figures showing that for the past three decades, the amount of beef Americans have been eating has been tumbling. Throughout the industrialized world, this trend continues. In my view, this industry tone deafness is partially to blame. Survey after survey shows Americans are becoming less accepting of chemical additives in their food (for decades, organics have been the fastest-growing segment of the food industry). Yet mainstream feedlot cattle nearly across the board continue to be raised with implanted hormones, and fed beta-agonists and antibiotics. At the same time, some of the meat industry's own research shows people who are reducing their consumption are doing so based on quality concerns. A recent survey on Meat + Poultry showed that people cutting back on red meat were also seeking red meat of a "higher quality" (although it's unclear exactly what is meant by the phrase). "Sixteen percent of Americans who say they are consuming less red meat are now consuming higher-quality red meat."[1] This tells the industry that if it wants to remain profitable, it needs to address consumer concerns. Still, the main initiatives we see from the beef industry continue to be about producing more meat for less cost, which surely *lowers* quality rather than raising it.

A specific example illustrating the way the cattle and beef industry typically responds to problems is the recent experience with beta-agonist drugs, like Zilmax. Manufactured by Merck, Zilmax is in a class of pharmaceuticals called beta-agonists, which function like steroids. They cause animals to bulk up quickly by converting feed to muscle production rather than fat. First introduced to the market in 2007, the drug was quickly widely adopted by feedlots. By 2012 it was estimated that 70 to 80 percent of the US cattle herd was being fed Zilmax (or a comparable drug called Optaflexx), generating $159 million in sales for Merck that year. Feedlots favor the drug because the animal gains more meat-producing muscle tissue on less feed.

However, as the drug became more widely used, reports of cattle lameness soon surfaced. Tyson Foods began to refuse cattle that had been fed Zilmax after it found that some cattle fed the drug were arriving at its slaughterhouses having difficulty walking. A Reuters investigation found some lame cattle fed Zilmax were missing hooves. Citing animal welfare concerns, and rejection by Asian markets, agribusiness company Cargill also announced it would no longer purchase animals fed Zilmax.[2] Astonishingly, none of this prompted the FDA to withdraw its approval of the drug (or do anything at all). Nonetheless, rejection by the big meat companies forced Merck, in September 2013, to voluntarily suspended sales of Zilmax in the United States and Canada. Note, though, Merck has indicated it does not believe there is any real problem with the drug and intends to compile more "research" to "prove" that the drug is actually safe. So we have probably not seen the last of it.

This was followed in 2014 by the release of research by a veterinary epidemiologist at Texas Tech University finding the incidence of premature death among cattle administered beta-agonists was 75 to 90 percent greater than cattle not administered the drugs.[3] World-famous animal welfare expert Temple Grandin has written and spoken publicly numerous times about the use of beta-agonists in cattle feed. She says cattle fed the drugs are "suffering." In one report Grandin wrote: "These observations indicate that there are severe welfare problems in some animals fed beta-agonists."[4] In an interview with National Public Radio, she said as many as one of every five cattle fed beta-agonists develop foot problems. Grandin seemed to implore the beef industry to fix the situation. "I've worked all my career to improve how animals are handled, and these animals are just suffering. It has to stop!"[5] China, Russia, and the European Union have already banned the use of beta-agonists in animal feed.

What's most telling is that despite this flood of negative information and press, and even trade repercussions, beta-agonists are still used nearly universally at feedlots. And beef trade publications defend their use. Shortly after Merck's withdrawal of Zilmax, *BEEF Magazine* ran a cover story about an agricultural college researcher whose work purports to show that, despite what you may have heard to the contrary, beta-agonists are perfectly safe. (Sigh.) This is what I call tone deafness.

Another example: feeding antibiotics. While the practice is far more prevalent at the confinement operations of swine and poultry, cattle feedlots feed a lot of antibiotics. They do this both to stimulate growth and to stave off diseases. Union of Concerned Scientists has estimated that cattle feedlots add about 3.7 million pounds of antibiotics to feed every year, with up to 55 percent of cattle being given subtherapeutic antibiotic doses.[6]

In recent decades, evidence has piled up about the dangers of antibiotics overuse in livestock. Due to human health risks, the US Centers for Disease Control and Prevention, World Health Organization, American Medical Association, and American Public Health Association have all taken positions against the practice of continually feeding antibiotics to farm animals. Every year for many years running, the US Congress has considered legislation outlawing the practice. Before his death in 2009, Senator Edward Kennedy was one of the lead sponsors of the US Senate's version of the proposed law. Nonetheless, year after year, the meat industry (including the beef sector) lobbies hard to defeat it. To this day, no federal legislation on the subject has passed.

In 2017, FDA issued regulations banning "medically important drugs"—those considered important to human health—from being used for growth promotion or feed efficiency. This is a step forward but does not go far enough. Routine addition of antibiotics to livestock feed is still allowed, and the practice remains widespread.

The meat industry claims in countries where the feeding of antibiotics has been banned, overall use of antibiotics has risen because there is more therapeutic use. This assertion has been repeatedly disproven by credible sources, including the Pew Commission on Industrial Farm Animal Production. The facts are clear: Banning subtherapeutic antibiotics feeding significantly lowers overall usage after the initial year. Despite FDA's action and more than a decade of effort by public health and environmental groups, it remains legal in the United States to add low doses of many antibiotics to the daily feed and water of cattle, other livestock, and poultry.

Contrast this with the way feeding antibiotics to livestock was handled in Sweden. In the early 1980s, Swedish consumers began expressing concerns about the overuse of antibiotics in livestock farming. Scientific evidence was emerging that feeding antibiotics to livestock was contributing to the rise of antibiotic-resistant diseases. The Swedish meat industry foresaw a crisis of confidence among its consuming public. It wanted to get ahead of the issue and to keep a level playing field for everyone raising livestock and poultry. So the meat industry *itself* lobbied the Swedish legislature for a law forbidding the continuous feeding of antibiotics to animals raised for food. With the support of the meat industry, Sweden passed the law in 1986. It was the first country in Europe to take this step. Following Sweden's lead, in 2006, a similar law went into effect for the entire European Union.

If you are part of the livestock or meat industry you have two choices: You can acknowledge legitimate issues and work to fix them, or you can engage in denials. As part of the cattle and beef industry, my hope is legitimate concerns will be recognized and solutions will be adopted by the industry itself. There is no better way to silence the critics.

Here are some specific things I think need work:

1. **Better grazing management.** Too much grazing is occurring without good planning and oversight. Well-managed grazing is not only environmentally beneficial, it is essential to functioning ecosystems; poorly managed grazing is damaging. Every farmer and rancher needs to get on board.
2. **Stop routinely killing primary predators.** Predators are essential to healthy ecosystems.[7] Ranchers need well-functioning ecosystems even more than the rest of society does. We must learn to deter predation and coexist with these essential creatures.
3. **Stop feeding drugs and other junk.** A shocking (and very unappetizing) list of drugs, including antibiotics and beta-agonists, and industrial by-products of all kinds are regularly fed to cattle, mostly at feedlots. This creates unhealthy animals, results in foods that may be unsafe for humans, and contaminates the waste streams coming from these operations. Nothing should be fed to cattle other than

pure feeds, comparable to what they'd eat in nature. Although I do not believe grains are inherently bad for cattle, I think they should be used sparingly because feeding them to cattle is a poor use of resources and leads to additional water and atmospheric pollution. All types and classes of cattle should be foraging to the greatest extent possible.

4. **Stop using hormones.** No growth hormones should be used on dairy or beef cattle. Period. As with beta-agonists, using hormones creates health and welfare concerns for cattle and leads to food that may be unsafe for humans. This worries a lot of consumers. It also limits the markets for US beef. The European Union banned the practice in 1981. All use of growth hormones should stop immediately.

5. **Stop putting calves in feedlots.** The optimal way to raise cattle is on grass for their entire lives. To the extent I find feedlots acceptable, I do not find them acceptable for young cattle. Health and welfare problems are magnified for young cattle.[8] If they are to go to feedlots, cattle should not go in until they are at least a year old, preferably 18 months.

6. **Stop slaughtering young cattle.** The routine slaughtering of young cattle (less than two years of age) is a relatively new practice in the United States. It has been abetted by the use of beta-agonists, hormones, and high-concentrate feeds, all practices that should be stopped (or, in the case of feeding concentrates, minimized). Cattle should, instead, be raised to full maturity before going to slaughter (a minimum of two years of age). This is a better use of resources and ensures better meat.

7. **Stop long-distance transport.** As every rancher knows, cattle do not lie down in truck transport. When they do, they are trampled to death. This is why it is essential cattle be shipped only moderate distances. According to Animal Welfare Approved standards, cattle transport should never exceed eight hours.[9] Shipping cattle long distances where they must stand longer is inhumane.

8. **Improve slaughter practices.** Cattle should be humanely handled at slaughter, period. It is not only the right thing to do, it has a strong correlation to meat quality and safety. The best way to ensure appropriate, low-stress handling is for the people who own the animals,

and are familiar to them, to do the handling themselves. This goes a long way toward ensuring the animals remain calm. Additionally, video cameras should be installed at all slaughterhouses to ensure humane handling of all animals at all times.

Reading this list in isolation, you might think it comes from a beef detractor. But I write it as a concerned member of the cattle industry. My hope is everyone who reads this list and who is involved with the raising, transporting, or slaughter of cattle will instead look at it as a *call to action*. If we wish to continue to exist, we can and must do better.

Why Eat Animals?

IN ADDITION TO WHAT'S ALREADY BEEN EXPLORED IN THIS BOOK, two major ethical questions surrounding beef consumption remain. One is whether it's morally acceptable to eat meat at all. The other is whether eating meat aggravates world hunger. I will address the first momentarily, and start with the latter.

The idea that beef contributes to global hunger undergirded the hugely popular and influential book *Diet for a Small Planet*. It argued, essentially, that the world's finite resources are stretched thin and will quickly be expended if a growing population of humans continues eating meat. The raising of livestock, especially cattle, the argument goes, is uniquely resource-intensive and cannot be morally justified in a world where (now) some 900 million people don't have enough to eat. In various forms, we continue to see this line of reasoning everywhere today, especially from vegan and environmental groups.

I agree meat and dairy consumption is out of balance. There's more than necessary in industrialized nations while there's not enough in developing countries, where malnutrition generally, and deficiencies of protein, iron, zinc, and vitamin B_{12}, specifically, are rampant. But there are many things wrong with the assertion eating meat adds to global malnutrition and starvation.

First I will point out something obvious: It can only be sensible to *quit* eating meat for this reason if doing so actually *aids* in relieving global hunger. For me, as a lawyer, this is self-evident. And yet, in over 20 years of reading various forms of the "livestock aggravates world hunger" argument, I have never seen anyone effectively demonstrate if you *stop* eating meat you will *help* world hunger. Rarely is such proof even attempted.

When distilled down to its essence, this is not really an argument that by refraining from eating meat you will help feed others. Instead it's more an endorsement of a principle of food equity: It's *unfair* to eat resource-intensive foods while others have insufficient food. But couldn't we just as easily argue we should refuse to drive cars or take flights because billions of people in the world can only afford to travel by bike or on foot? Should we refuse to use air-conditioning while others live through debilitating heat and cannot cool themselves? These are interesting ethical questions. And any person is certainly free to take one or all of these positions for ethical reasons. But it should be recognized as a symbolic moral stance rather than an altruistic act. None of those actions would actually help a single person in need.

At no time in recent history has global hunger actually been the result of an inadequate global food supply. According to food and nutrition expert Dr. Marion Nestle, the world produces some 3,800 calories per day for every man, woman, and child on earth. This amount, she notes, is almost double what's necessary for adequate nourishment.[1] For the last four decades, per capita food production has actually grown at a pace 16 percent faster than the world's population.[2] In his book *World Food Security*, Dr. Martin M. McLaughlin, who has worked on food security issues and taught university courses on the subject for decades, makes plain that world hunger has very little connection with the quantity of food the globe produces.[3] Poverty, not food shortage, is the real issue, says McLaughlin. "Hunger . . . is a political and social problem," he writes. "It is a problem of access to food supplies, of distribution, and of entitlement."[4] Livestock assist people in feeding themselves and are an essential income source that allows people to purchase food staples.

Livestock farming is clearly not the province of the rich—very far from it. An article in the respected journal *Nature* points out one billion of the world's poorest people depend on livestock for their survival.[5] A 2013 FAO report similarly states: "Hundreds of millions of pastoralists and smallholders depend on livestock for their daily survival and extra income and food."[6] In developing countries many poor families, including those who own no land, have a cow or goat or some chickens. The eggs, milk, and meat make up an irreplaceable component of their income and food. "Almost every smallholder farming family in a developing country owns livestock, whether chickens, rabbits, sheep, goats, pigs, cows, buffaloes, donkeys, horses, yaks, llamas, or camels," states a 2014 article by an agriculture official from India.[7] "Livestock development benefits poor rural families, many of them engaged in farming but not owning land."

Livestock keeping offers several key advantages not afforded by plant crops for gaining food and financial security. Most fundamentally, you need to own or rent land to grow crops. This is not the case with livestock, which can be led to common grazing areas and/or fed gathered forages, food scraps, and other edibles that don't require having control over a piece of land. And in contrast with crop farming, which produces sporadic, seasonal, perishable products, livestock is an asset that can be maintained for short or long periods of time then quickly converted to food or cash when needed. This has been the case since people began keeping farm animals thousands of years ago. As writer and farmer Simon Fairlie points out, "a main role of animal husbandry has been to provide food security: 'The purpose of domestication was to secure animal protein reserves and to have animals serve as living food conserves.'"[8] This is why livestock are sometimes referred to as "an ATM for poor farmers." Everywhere in the world, the flexible nature of animal keeping has always been among its primary benefits. Harold McGee writes: "Livestock not only transformed inedible grass and scraps into nutritious meat, but constituted a walking larder, a store of concentrated nourishment that could be harvested whenever it was needed."[9]

In many parts of the world, farm animals raised for meat and milk also provide invaluable labor and transportation services. Oxen were once common in the United States, and continue to perform essential services on some smaller US farms. They still pull plows and carts on a large portion of the globe. These are all attributes utterly unique to animal keeping.

Females, who make up most of the animal tenders in the developing world, are often livestock's greatest beneficiaries.[10] Animals provide women reliable income and protein-rich foods for their own families, both available on an as-needed basis.

Livestock are also an excellent complement to plant farming. "Half the world's food comes from farms that raise both," science journal *Nature* points out. "Animals pull ploughs and carts, and their manure fertilizes crops, which supply post-harvest residues to livestock."[11] In his book *Feeding People Is Easy*, veteran British science reporter Colin Tudge, who has traveled the world extensively and reported for decades on food and agriculture, states "pastoral farming is very important indeed." He declares: "The oft-bruited generalization—that we could most easily feed the world if everyone was vegetarian—is simply not true." Among the reasons he points to are, "there is no system of all-plant agriculture that could not be made more efficient, in biological terms, by adding in a few livestock, provided they are the right kind, and are kept in the right numbers, in the right ways."[12]

That farm animals are lifeblood for hundreds of millions of the world's poor became much clearer to me after I attended an international gathering of smaller-scale livestock keepers. The Livestock Futures Conference in Bonn, Germany, hosted by the nonprofit League for Pastoral Peoples, convened 70 livestock keepers and researchers from 16 different countries and several continents. Among those in attendance were people involved in camel herding from Pakistan, cattle herding from Uganda, sheep herding from Germany, and goat herding from Argentina. (Henning Steinfeld, lead author of the FAO's *Livestock's Long Shadow* report, was there, too.) I was honored to have been invited to speak about issues facing livestock farmers and ranchers in the United States. The organization's founder, a remarkable German woman named Dr. Ilse Köhler-Rollefson, a doctor of anthropology and veterinary medicine who has lived for extended periods among camel-herding people in India, has labored for years to raise the profile of smaller-scale livestock keepers before the world's policy organizations. On the global level, just as on a national one, small-scale farmers and livestock herders are often overlooked, as the biggest players are given greatest consideration.

The Livestock Futures Conference highlighted the social and ecological contributions of small operations, contrasting them with environmental and societal damage caused by industrialized livestock production. Research and first-person testimony described wide-ranging benefits to smaller keepers, including ecological and climate protection, cultural preservation, tourism, and support of local labor markets. Presenters showed that in many of the globe's arid regions, places where land cannot be put to other uses, the small-scale livestock sector is responsible for the largest share of animal production. All told these smaller farmers make up 30 percent of world production of animal-based foods. Despite their essential contributions to food security, small farmers and herders are still being ignored in national and international policy.[13] The conference was part of a multiyear strategy by the league and its allies to change that.

Even if we accept without proof that eliminating livestock would lead to more food for the poor (which I definitely do not), the notion relies on exceedingly fuzzy math. Let's look at some of the problems. First off, it assumes grain currently grown and fed to livestock would still be grown and would somehow end up being made available to the world's poor. This defies logic and ignores realities on the ground. If livestock were no longer generating a demand for grain, why would farmers and agribusiness companies continue growing it? They would not. Like anyone engaged in the production of any commodity, they would adjust, and farmers would shift to growing crops for which there was the most demand. More sugar beets, perhaps? In the alternative, they might convert the land to other, non-agricultural uses. Under either scenario, global hunger is not reduced.

For another thing, the high-quality nutrition of meat and milk would be difficult to replace with foods from plants, especially for people in the developing world and children everywhere. Substituting plants for all animal-based foods would take far more than a one-to-one pound-for-pound or calorie-for-calorie replacement by grains and soy. The Pimentels say this very clearly: "Animal proteins contain the eight essential amino acids in optimal amounts and in forms utilizable by humans for protein synthesis. For this reason, animal proteins are considered high quality proteins. By comparison, plant proteins contain lesser amounts of some of the essential amino acids and are judged to be lower in nutritional quality than animal sources."[14] This is especially important for children, they note, whose rapidly developing bodies particularly benefit from nutrient-dense foods. "Another advantage of animal products over plant products as food for humans, especially children, is the greater concentration of food energy per unit of weight compared with plant material. For example, . . . beef has three times as much food energy per unit of weight as sweet corn."[15] Nutritionally, animal-based foods are more important to the world's poor than any other foods. For the one billion of them who raise their own livestock, it's critical for feeding themselves. Eliminating animals from the food system would likely make the world's hungry *more* food-insecure, not less, and more dependent on government assistance.

Moreover, while world grain use for livestock is significant (too high, in my view), it's far less than what people generally assume. Cattle in the developing world are usually fed little or no grain. In the United States, the breeding herds of beef cattle (about 30 million animals) are generally also maintained with little or no grain. Nearly all steers and heifers raised for beef in the United States are raised on mother's milk and range or pasture, then fed grains only in the latter portion of their lives.

Even for cattle fed grain in industrialized countries (both beef and dairy types), a large portion of their diet still comes from substances humans can't eat—forage or farm by-products (like straw or rice bran). The *Nature* article notes that around 70 percent of grains used by developed countries are fed to animals, with 40 percent of such feed going to ruminants, mainly cattle.[16] I believe this amount can and should be pushed down significantly. Even so, the same article points out, much livestock feed in developed countries consists of plant matter inedible to humans. "Even where large quantities of cereals are consumed by ruminants, up to 60% of their diet comes from high-fibre feed that humans cannot digest . . . In the European Union, more than 95% of milk comes from animals fed on grass, hay and silage, supplemented with cereals."[17]

In the developed world, a portion of milk and beef comes from animals raised entirely on grass. Some American beef cattle, our own among them, are raised from birth to slaughter without ever eating grain. "Cattle in New Zealand's exemplary dairy industry obtain 90% of their overall nutrition by grazing pasture."[18] New Zealand's dairy industry and grass-fed cattle ranchers in many parts of the world, including Canada, the U.K., Australia, and the United States, demonstrate the feasibility of a worldwide transition back toward forage-based diets for ruminants, including both beef and dairy cattle. Probably, this shift would mean that some amount of production would be lost. But that reduced volume would be more than offset by the overall benefits to the environment, animal health and welfare, and human health.

Energy use for cattle fed from their own foraging is negligible. It's so low that grass-fed beef is actually less energy-intensive than grain. The Pimentels point out in *Food, Energy, and Society* that crop cultivation adds significantly to the energy use of grain-fed livestock systems: "[C]attle grazed on pastures use considerably less energy than grain-fed cattle."[19]

The textbook quantifies energy inputs for grass-fed beef compared with grain-fed as follows: "Current yield of beef protein from productive pastures is about 66 kilograms per hectare, while the energy input per kilogram of animal protein produced is 3,500 kilocalories. Therefore, animal protein production on good pastures is less expensive in terms of fossil energy inputs than grain protein production."[20]

The other part of the fuzzy math problem is about land where livestock actually graze. The "stop eating meat to reduce world hunger" notion assumes if farm animals disappeared, a significant portion of grazing land could (and would) be used to raise food directly for humans instead. This is wrong on several levels. First, there is no reason to believe land currently used for grazing would be used to grow food for the world's poor. Whoever owns or controls the land would find other, more profitable uses. In the alternative (such as federal lands grazed by US cattle), the land would simply cease to be used for food production. Thus, removing livestock would not free up land for plant farming, as people making this argument generally assume.

Second, and this point is critical, the vast majority of the world's grazing takes place on land that *cannot* be used to grow crops. Earth sciences professor David Montgomery succinctly explains: "Sheep and cattle turn parts of plants we can't eat into milk and meat."[21] *Food, Energy, and Society* notes the prevalence around the world of livestock raised on "free energy sources." These include forage growing along paths and other "interstitial spaces" that would not be used for crops or other purposes, and straw left after harvest of rice or similar grain crops, which can be fed to animals.[22]

The college textbook *Soil and Water Conservation* defines *rangelands* as "soil on which the native vegetation is predominantly grasses, grass-like plants, forbs or shrubs suitable for grazing or browsing." It notes, "Nearly half of the land on Earth can be classed as rangeland," and says, "Most of it is either unsuitable or of low quality for use as tilled cropland because it includes steep areas, shallow and/or stony soils, or dry and/or cold climates."[23]

The National Sustainable Agriculture Information Service provides the following useful explanation of the unique role of grasslands and ruminants like cattle in the global food system:

[Grassland] ecosystems are naturally able to capture sunlight and convert it into food energy for plants . . . [M]ost of the land in the U.S., and indeed in most countries of the world, is not tillable and is considered rangeland, forest, or desert. These ecosystems can be very productive from a plant biomass perspective, but since they are generally non-farmable, *the plants they produce (grasses, forbs, shrubs, trees) are not readily usable (from a digestive standpoint) by humans.*

However, grassland ecosystems (both rangeland and temperate grasslands) produce plant materials that are highly digestible to ruminant animals . . . Grazing of native and introduced forages on grasslands and rangeland thus is a very efficient way of converting otherwise non-digestible energy into forms available for human use: milk, meat, wool and other fibers, and hide.[24]

The bottom line is this: The miraculous transformation of sunlight into human food via grazing animals is mostly occurring in areas that cannot be used to cultivate crops. This crucial fact is nearly always ignored by (or perhaps unknown to) beef's critics. Colin Tudge uses his book *Feeding People Is Easy* to raise awareness on this essential point. He writes:

In many parts of the world, at least in some seasons, it is very difficult to raise crops at all. Arable is all but impossible when the land is too high, steep, cold, or wet or if it rains too much in the season when the grain should be ripening . . . *[A]nimals of one kind or another muddle through anywhere—living as camels and goats may do on the most meager of leaves that poke through between the thorns of desert trees, or as reindeer do on lichen, or as long-wooled sheep and shaggy cattle do in British hills on the coarse grasses that grow between the heather; and in times of drought or the depths of winter there may be nothing to eat at all except for beasts that fattened in better times.*[25]

A recent grasslands university textbook reinforces Tudge's observations with respect to rangelands. It states: "Because they typically experience low and unpredictable rainfall and often have associated low soil fertility, rangelands generally cannot sustain crop agriculture without irrigation."[26] Making good use of these lands is and has always been the very special role of the world's herbivores, the text then notes. They convert "low quality fibrous plants into products such as meat, milk, and blood that humans can readily digest." For this very reason, the text continues, "harvesting products from herbivores has been a defining element in the relationship between humans and rangelands worldwide for millennia."[27]

One of many world examples of people using livestock as rangeland converters are the Dodos of northeast Uganda. They feed their cattle no grain, only forages unsuitable for human consumption, and raise them without fossil fuels. The Dodo tribe illustrates the crucial and versatile role livestock can play for humans. *Food, Energy, and Society* summarizes the benefits: "First, the livestock effectively convert forage growing in the marginal habitat into food suitable for humans. Second, herds serve as stored food resources. Third, the cattle can be traded for sorghum grain [for human consumption] during years of inadequate rainfall and poor crop yields."[28] These are precisely the properties that make livestock irreplaceable to people throughout the world.

In the United States, since long before the arrival of humans, a large portion of ground, especially in the West, has been unsuitable for crop cultivation. Some of this land is arid or semi-arid; rains may be insufficient or fall only at the wrong time of year for crop cultivation, or the topography is too hilly, or too rocky.

I'm a child of the Midwest, so the climate and topography of the West were long unfamiliar to me. But now, having resided for the past 17 years in Northern Coastal California on just this sort of land, I understand such limitations much better than I once did. I grasp implicitly how windiness, dry, cool summers, and steep, rough terrain are all conditions that are fine, even ideal, for grass and livestock, but render crop growing impossible.

"The [US] pastureland and rangeland are marginal in terms of productivity because there is too little rainfall for crop production," note the Pimentels.[29] These areas are where the vast majority of America's cattle are located. According to the US Beef Board, 85 percent of the land grazed by cattle in the United States is land that cannot be farmed.[30] This precise number, since it comes from the beef industry itself, obviously should be taken with a grain of salt. But it suggests cattle grazing in the United States is largely occurring on non-farmable lands. Another source, a credible recent university textbook, states that in California,

57 million acres, almost 60 percent of the land, is characterized as range-land. Only about 34 million acres of it is actually grazed.[31]

Yet in California, as elsewhere in the United States, rangelands continue to be chipped away by more intensive land uses. Foothill rangelands, especially, are being converted to wine-grape growing, housing, and urban developments. A university grasslands textbook states: "[T]hroughout most of California, range has given way to other, generally higher value but also more intensive, land uses."[32]

Yet on a more optimistic note, the same textbook holds out hope for a growing recognition of rangelands' societal and ecological value:

As range ecosystem services other than livestock production become increasingly valued by society, the additional benefits that we derive directly from primary production and the soil system have gained greater recognition: provision of irrigation and drinking water, recreational opportunities, wildlife habitat, open space/viewshed, rural lifestyle, biodiversity, and carbon storage. It might even be argued that the most important role that primary consumers play in 21st century California is not providing the traditional livestock products of meat, milk, and fiber, but rather acting as a bulwark against the conversion of range into housing developments, vineyards, and other more intensive land-uses that do not provide the multiple ecosystem services bestowed by range ecosystems.[33]

The presence of grazing animals on such non-farmable lands enriches the world's food supply regardless of the efficiency with which the animal converts the feed to flesh and milk. Because they're living from inedible vegetation, the oft-quoted statistics about the "inefficiency" of cattle converting feed to flesh are irrelevant. "However efficient the conversion ratio of any given animal may be," Simon Fairlie points out, "if it is grazing entirely on land which could not otherwise be used for arable production or some other highly productive activity, then it cannot be said to be detracting from the sum quantity of nutrients available to the people of the world, but adding to them."[34] Moreover, Fairlie notes, where animals graze on land unsuitable for crop cultivation, they are "relieving pressure on arable land, and helping to retrieve otherwise inaccessible nutrients and bring them within the food chain."[35]

To this I would add livestock that graze on cover crops and those fed farm by-products. A large portion of the world's grazing animals, including in the United States, are raised in such systems. These methods both increase the biological vibrancy of farmed land and improve the food system's efficiency. They create food for humans using only feed sources not directly usable as human nourishment.

I hope this discussion eases the mind of any reader who has hesitated to eat beef based on concerns about world hunger. In most of the world, cattle are fed little or no grain and are raised on non-farmable lands. For Americans and other people in the developed world, where grain is used as part of cattle feed, we have the choice to seek out and buy beef and dairy products from animals raised on forages rather than grains and soy. As described earlier, there are compelling human health and animal health and welfare reasons to do so. When we choose grass-based foods we also help maintain our nation's grasslands, which are the most environmentally beneficial of all agricultural lands.

The other aspect of the moral question about eating beef is whether it's acceptable for humans to eat meat at all. In answering this, I'd like to rely less on data and statistics and bring some of my own personal experiences to bear. It's been 20 years since I began working on farm-related issues for Waterkeeper. Although my job there was focused on mitigating pollution, to me the animals were equally important. Other environmental groups were advocating addressing industrialization with improved waste containment or treatment. But I found those approaches far too narrow. That turns a blind eye to factory farming's greatest evil—animal cruelty. Even worse, by endorsing pollution reduction that failed to improve farm animals' lives, these groups were effectively further entrenching the current cruel system.

Fortunately, my boss felt exactly as I did. Bobby Kennedy Jr. has cared passionately since childhood about creatures from sow bugs to blue whales. He heartily supported my advocating for farming that was ecologically sound *and* provided animals good lives. Having gotten a close, personal view of the brutality of industrial production, we, and our colleague Rick Dove, all felt morally obligated to help farm animal welfare.

In the years since, my writings and speeches have often touched on the ethics of meat production. Until a few years ago, though, I never focused on whether or not it is ethical to eat meat at all. Going back to ancient times, there have been pro and con vegetarian debates. The debate continues to rage. I always preferred to focus my energy on the question of *how* we are raising farm animals, not *whether* we should.

Then an ecological journal invited me to write an essay arguing that an environmentalist need not refrain from eating meat. I was still a vegetarian at the time but decided to accept. My essay made the case that animals are a crucial part of all well-functioning ecosystems. Following its publication, I was invited to participate in several live debates about the ethics of meat eating, two of which I accepted. On both occasions (still a vegetarian at the time), I represented the "pro-meat" position. I later wrote a piece for TheAtlantic.com titled, "Can Meat Eaters Also Be Environmentalists?"[36] Even though I never sought to focus my energies on the question of whether people should eat meat, it seemed important to refute the increasingly prevalent notion that people who cared about the environment should avoid it. I was a vegetarian meat advocate.

That first piece, titled "Animals Are Essential to Sustainable Food," noted that today's debate over meat lacks necessary nuance and honesty. It's characterized by polarizing, oversimplified rhetoric, pitting an implacable, defensive agribusiness in one corner against intractable, aggressive vegan activists on the other. "[The abolitionist vegans'] fervent advocacy echoes prohibitionists at the dawn of the twentieth century," I wrote, "some of whom attacked apple trees with axes because they were the source of hard cider. Like the prohibitionists, activists against meat are fueled by the excesses of the day."[37] Factory farms, not animal farming, are the real problem, I urged. And industrial methods have fostered a growing disillusionment with the meat industry among broad swaths of the American public, well beyond vegan and vegetarian circles.

I fully agree that today's industrial methods for raising farm animals are indefensible. Everyone should join in rejecting them. Having seen it for myself, I have no qualms about calling industrialized animal production a routinized form of animal torture. Prohibitionists seem absurd for assaulting innocent apple trees with axes. But like a lot of discussion over the ethics of meat eating, the real failure was that it attacked the wrong villain. Alcohol, not apple trees, was their enemy. Industrial animal production should be vilified. Animal farming should not.

I'm not interested in the debate over meat because I want to encourage its consumption. What I long for is some nuance and thoughtfulness in the discussion. It's a complex and emotionally charged issue with many subtleties. Polarized camps lobbing accusations at each other don't move us forward. They hinder progress. Building an ecologically resilient and humane food system is far more important to me than whether or not someone is eating meat.

As I see it, the real issue is whether we humans are living up to our responsibilities as good stewards of animals and of the earth. Michael Pollan and others have proposed that animals "chose" domestication as a "bargain" with humanity. (I put the words *chose* and *bargain* in quotes because obviously no individual wild animal ever made a conscious decision its species should be domesticated.) Domestication likely happened gradually over many generations as some animals found advantages to having a certain amount of human contact. Humans "agreed" (again, in quotes, because the bargain was implicit) to provide essentials to animals—food, shelter from the elements, and protection from predators being foremost among them—in exchange for the animals providing humans food in the form of eggs, milk, and meat. (With dogs, the terms of the bargain were quite different. For being provided protection and nourishment, dogs exchanged assistance in hunting, early warning, and self-defense.)[38] However, it's reasonable to assume, as well, that animals would never have opted for such an arrangement if torture had been part of the deal. Stated simply: Factory farms are violating humans' age-old contract with domesticated animals.

For American farm animals, consideration of the individual animal's dignity was forsaken mid-20th century. *Husbandry* implies skill, care, and consideration. As confinement operations were becoming the norm, agricultural colleges changed the appellation of their "animal husbandry" departments to "animal science" departments. The word change embodies the shift in mind-set.

Agribusiness has long defended its methods by pointing to their prevalence. But we all know in our hearts that just because something is widespread doesn't mean it's acceptable, let alone right. Factory farms are undeniably inhumane. The worst practices are narrow metal cages for pregnant sows, wooden crates for veal calves, and wire cages for egg-laying hens. But beyond that, the everyday workings of industrial facilities utterly fail at providing animals decent lives. Continually keeping animals in foul-smelling, cramped quarters, on hard floors, depriving them of all pleasures and basic necessities like exercise, fresh air, sunshine, and a soft place to lie down, cannot be called humane. Whatever rationale is offered for these practices—"efficiency," "cost of production," "affordable food," "feeding the world"—these systems remain morally indefensible.

Grazing animals, especially those raised for meat (rather than milk), have fared better than others. Their unique capacity to sustain themselves on grass has been their saving grace. Keeping them outdoors on naturally occurring vegetation is often the most economical way to raise them. Nearly all cattle, including dairy heifers, spend their early life on grass. Once mature, most (although not all) dairy cows with modern genetics, bred for high-volume milk production, are confined. They are fed concentrates, the only way to achieve their full genetic potential for voluminous milk production. Their time on grass is then over.

Beef cattle have it better. Those raised for meat (not kept for breeding) typically go to a feedlot sometime before one year of age. Even there, they are out in the open air and have the benefit of soft ground for lying and standing. The breeding beef animals (mother cow herds and bulls) are the most fortunate. Generally, they spend their entire lives on pastures or rangeland, having a daily existence not unlike that of their wild ancestors. Because a cattle ranch's success depends on mother cows being able to survive and give birth without human assistance, beef cattle have long been selected for heartiness and good calving ability. In this, the interests of the animal and those of the rancher perfectly coincide. These traits help rather than hamper the individual animal's quality of life.

Beef cattle are so much better off than other farm animals that, as mentioned earlier, an animal-activist friend says he considers it better to eat beef than eggs. Soon after my work became focused on agriculture I, too, became persuaded that cattle raised for beef are the luckiest of the animals raised for food.

From a dining perspective, I'd much rather eat something derived from an animal that spent its life exercising, breathing fresh air, and grazing meadows. Food from animals cooped up in crowded, stinking metal warehouses is not appetizing. Why would I want to eat foods that originated from places I would never want to visit?

On top of those issues, there is the amount of meat per animal to be considered. Previously I noted that taking the life of a single steer provides more than ample meat for two families for an entire year. The industry average is 475 pounds of beef per carcass. In stark contrast, the industry average weight for a whole chicken carcass is 4 pounds, with about 70 percent of that being meat. A chicken yields less than 3 pounds of meat. You have to kill more than 150 chickens to get as much meat as you receive from a single steer. It's hard to know how to compare the morality of killing 150 chickens versus killing one steer. But for me, it was among the reasons I began favoring beef over other animal-based foods even before first stepping foot on my husband's ranch.

Even during my 30-plus years as a vegetarian, I never could embrace the view that eating meat is ethically wrong. We can debate until the cows come home things like evolution of teeth, digestive tracts, and other aspects of human physiology to make a case that humans are either "intended" to eat meat or not.[39] But to my biology-trained brain, those points were never very persuasive. Humans may have evolved from herbivores if you go back to a certain moment in time. But the ancestors of those animals were omnivores and carnivores. (You know what a shrew eats? Insects, slugs, spiders, worms, amphibians, and small rodents.)

And it's now believed that human ancestors began eating meat at least 2.6 million years ago, with a major uptick in meat consumption occurring around 1.5 million years ago. Clearly, a great deal of evolving has taken place in those millions of years. Any way you slice it, modern humans come from a very long line of meat eaters.

My view boils down to this: Humans are animals belonging to a complex food web. The web includes animals eating plants, animals eating other animals, and even plants eating animals. As this book has been at pains to portray, all life starts from the earth and returns to the earth. The bodies of all plants and animals nourish future generations of plants and animals in an endless cycle of growth, decay, and regeneration. Ashes to ashes and dust to dust. To me, something so fundamental to the functioning of nature cannot be regarded as morally wrong.

As I've learned how animals in the food system live, it has also brought me to the realization, ethically speaking, that there is little difference between eating meat and refraining from meat if you are still eating dairy and eggs. Dairy cows all become beef eventually. And, as just discussed, most dairy cows don't live as well as beef cattle. Most egg-laying hens spend their lives crammed into wire cages (so-called battery cages), then are usually unceremoniously vacuumed up for use in things like canned chicken soup and pet food. To feel any sort of moral superiority for not eating beef while eating dairy products and eggs is absurd. Once that point became clear to me, it bridged any ethical distance I might have once felt between myself and my husband's vocation.

Still, I initially had some uneasiness about moving to a cattle ranch. Sure, I supported ranching *in principle*. But I wasn't sure I'd feel comfortable living day after day in the midst of an active operation. Would I find it upsetting to be surrounded by animals I knew would one day be sent to slaughter? And more to the point, would it make me feel guilty knowing our living came from their deaths? I vaguely figured I'd keep myself at arm's length to avoid any potential discomfort.

What happened instead was unexpected. For exercise and to take in the area's natural beauty, I took long walks nearly every day through our land. Almost by accident, I began regularly spending time in the company of our cattle. I saw our mother cows ambling as they ate, socializing with their sister herd members, congregating around the water trough, calling their babies, licking one another's necks. I watched calves frolicking in the grass, racing around after one another at twilight, running to their mothers for comfort and a long, warm drink of milk. I saw the bulls and cows nuzzling each other in courtship before and after mating. It was easy to see these animals had lives well worth living. The more I meandered our meadows and sat on fences observing, the more I valued and appreciated what was happening in my midst. Everywhere I looked I saw animals living well and well-cared-for lands.

After a few months, I told Bill I wanted to learn to do everything on the ranch. This surprised him a bit. He, too, had expected his vegetarian wife to want to keep a certain distance from ranching operations. But he readily agreed. Soon I became the wide-eyed, unskilled, displaced city-dweller ranch hand. I did my best to help him and our ranch manager with whatever needed to be done each day. Occasionally, this involved fixing a fence or a water trough (although my role was usually tantamount to holding the tools). In the dry season, it meant bringing hay to the cattle. Daily, it meant taking my pocket-sized notebook and walking or riding on horseback through the herd, making sure each animal was healthy and accounted for. Gradually, I came to know each herd member individually. I learned how to do things, and was given more responsibility.

After a couple of years, and for several years following, I became the primary labor on our ranch. I loved doing physical work, outside, especially being among the animals. I liked the daily challenge of problem solving. It was especially rewarding when I could be useful to the animals. Sometimes I'd reunite a mother with her new calf who'd slid beneath a fence. Occasionally, we'd successfully graft a spare twin calf onto a mother cow with a stillborn. These physically and emotionally intense tasks were deeply rewarding moments, etched in my memory and my soul.

A lot of my time was spent watching and interacting with wildlife. I became daily witness to nature's beauty, its force, and its cycles. A mother bobcat stealthily stalking her prey; an osprey cruising silently overhead with a fish clasped in its talons; a wake of vultures feasting hungrily on a deer carcass. Grasses and wildflowers sprouting, blooming, drying, dropping their seeds, dying back. Life-giving rains arriving to our relief. The cycle beginning anew.

These experiences reinforced how all life is connected. Living in Manhattan, as I did for five years before moving to the ranch, I walked on concrete in high heels, surrounded by metal, brick, glass, and the cacophony of car horns, passing airplanes, and subway cars. It had been easy to see myself as separate from the natural world. Working every day out on the ranch rebuilt the strong cords I'd felt with nature most of my life.

My ranch work also taught me that even the most conscientious agriculture is a major disturbance. No matter how well done, farms and ranches will invariably have profound effects on all shapes and sizes of wild creatures, plants, fungi, soils, and water. We should not expect people growing our food to have zero environmental impact. This is unattainable. Instead, the goal should be food production in harmony with nature.

In his brilliant book *Call of the Reed Warbler*, Charles Massy, Australian sheep farmer and PhD ecologist, advises aiming to farm in ways that restore proper landscape function. As we work the land to generate food, we must constantly pay attention, learning to understand earth's systems. Rather than fighting them, human-built systems should follow these natural patterns. When we do so, we can restore the original functioning of our landscapes, bringing back vitality, life, and healthfulness, Massy argues.

Consider crop cultivation. Especially if done using typical modern farm machinery, it represents an enormous disruption to naturally occurring vegetation and animals. Plows scrape, cut, and chop up the earth's surface and whatever is growing there. They tear up complex communities of everything from symbiotic microorganisms to rabbits, voles, and snakes. In the developed world, most plows are dragged behind enormous, heavy tractors. This is followed later in the season by huge harvesters that crush and shred every plant and animal in their path. These machines bring Armageddon for billions of soil-dwelling creatures along with every form of wildlife residing in or on the ground. There's no avoiding the reality that large-scale crop farming is the most disruptive of all agricultural acts.

At the start of his beautifully photographed and eloquently written *The River Cottage Meat Book*, British chef Hugh Fearnley-Whittingstall poses these questions: "Why do we eat meat? And is it right, morally, that we do?"[40] In answering, he starts by noting no matter what we do, it's a fallacy "to maintain that we can live in complete harmony with the rest of animal kind." The actions of all animals, he points out, whether intentionally or not, will affect others. "The undeniable fact is that any species' pursuit of its interests will always have an impact on the rest of the planet's life—the fox impacting on the chicken population, the flea on the cat, the beaver on the forest, and the sheep on the grass."[41]

I agree with Fearnley-Whittingstall's implicit suggestion no human endeavors, and certainly not farming, can be done without impacts on other life-forms. The notion we can "eat cruelty-free" by avoiding foods derived from animals is a fiction. If we really look at agriculture, especially crop farming used to produce plant-based foods, it becomes quickly apparent there is injury all along the way. Cultivating soy, grains, fruits, and vegetables unavoidably maim and kill animals and alter habitat for literally billions of creatures.

Instead of seeking an unattainable absence of cruelty, I'm looking for agriculture that respects all life and follows nature's model. In the end, answering the question: *Am I eating food derived from an animal?* tells you very little. It says nothing about what impact production of that food had on the animals, plants, and fungi coexisting in that farming ecosystem. All farming, and especially crop farming, kills animals of various shapes and sizes. The more meaningful question is: *Has this food been produced as nature functions?* For me, it's clear that sort of farming involves animals.

Sir Albert Howard, considered a godfather of modern organic farming, viewed animals as inextricably linked to ecologically sound food production. He called animals "our farming partners." Howard said: "In Nature animals and plants lead an interlocked existence. The connection could not be closer, more permanent, or more crucial. We can observe this partnership in operation in the forest, in the prairie, in marshes, streams, rivers, lakes, and the ocean."[42] Howard's point, I believe, is that nature's way of converting water and sunlight to energy, in the form of food, is a complex, multi-stage process in which no part exists in isolation. Each component—every ray of sunshine, every drop of water, every clump of soil, every plant, every insect, every grazing animal—has many and varied roles and effects. The more farming systems reflect such complexity, the more ecological they are.

One of the most thorough and thoughtful explorations I've seen of why we should farm with animals is in Simon Fairlie's book *Meat: A Benign Extravagance*. Fairlie's perspective is informed by his diverse experiences, which include environmental writer, farmer, and former vegetarian. He, too, believes farming ecologically takes animals. His book describes an optimal food system where cattle and other grazing animals convert non-tillable lands into contributors to the food system while omnivorous animals, like pigs and chickens, make good use of farm by-products and food scraps. He also emphasizes the high value of manure for post-fossil-fuel fertility.

Fairlie also points out that "wherever there is livestock there is the opportunity to garner and concentrate fertility from the wilder environment . . ."[43] This is especially the case for animals that graze. More efficiently than machines, Fairlie notes, grazing animals "move nutrients from where they are not needed to where they are required." This is particularly important with phosphorus. Unlike nitrogen, phosphorous cannot be obtained from the air. "[P]lants cannot extract phosphorus

from the atmosphere, so the role of animals in importing surplus phosphorus from outlying areas could be crucial," he argues.[44]

Fairlie also points out domesticated grazing animals have taken
the place of wild herbivores in keeping a balance of open spaces with
wooded areas. Light with shade, as he puts it. The balance, he suggests,
would massively shift without the presence of grazing animals:

> *Only livestock can engineer the balance that any society seeks between
> the realm of light and the realm of shade on any scale beyond the arable.
> Even in a full-blooded fossil fuel economy, JCBs, timber harvesters and
> other wheeled monsters are fighting a losing battle with nature unless
> they enlist the help of quadrupeds. Where livestock are allowed to roam
> they bring grass, and where they are excluded trees grow, and it is a
> relatively effortless matter for humans to calibrate their performance
> to our will or whim.*[45]

In a related vein, the university textbook *Soil and Water Conservation* describes the value of livestock for managing fire risks on a broad,
landscape scale. "When livestock graze these fire lanes, forest roads, and
forestlands in general, they reduce the fire hazard by removing vegetation
that is flammable when dry."[46] The way that cattle manage vegetation,
holding back the spread of woody plants and keeping open spaces open,
is something few of us pause to consider or appreciate. The look and
functionality of our landscapes would be radically different without them,
and in many ways for the worse. Humans really do need grazing animals.

We hear a great deal about the planet becoming crowded and harder
to feed. All too often it's suggested livestock are part of the problem.
Because it doesn't fit neatly into the advocacy narrative for vegans and
environmentalists, though, we rarely hear about crop cultivation destroying agricultural land. That kind of farming is actually the greatest threat
to humans' ability to feed themselves in the future. In *Dirt: The Erosion of
Civilizations*, David Montgomery tells the history of societies failing to
properly steward their soils, to the point where their lands are no longer
usable for crop cultivation. This continues at an alarming pace today. *Food,
Energy, and Society* says, worldwide, more than 50 million acres of agricultural land is abandoned annually because of soil erosion and salinization

from irrigating crops. "During the past 40 years, about 30% of total world arable land has been abandoned because it is no longer productive," the book notes. It is estimated about half of the land currently under cultivation will be unsuitable for food production by the middle of the 21st century.[47]

People are rightly concerned about adequate future food supplies. But they should turn their attention to the imminent crisis of industrial farming, and on how animals are managed on the land, rather than on the number of livestock, which tells nothing about whether farming systems are ravaging our landscapes or regenerating them. Properly managed grazing animals are an important part of the solution to feeding the world now and in the future.

I may once have believed that if I followed a vegetarian diet, nothing would have to die for my meals. I now see how wrong I was. The more I've gotten to know farming, the more that seems a gross oversimplification. The earth has changed constantly over its hundreds of millions of years. But it was never assaulted on such a broad scale until humans started crop agriculture. The effect that was greatly amplified with mechanization. Pastoral animal keeping, however, mimics the functions of wild herds that covered the earth for millions of years. The impact of animal herds is a *familiar* disturbance to the earth. It's an impact soil organisms, plants, and animals not only can tolerate but actually *need*. Vegetation will be pruned and stepped on, as it has always been, allowing for a diversity of plants, and for perennialism. Compared with other ways of producing food, the keeping of grazing livestock, when done appropriately, is the most environmentally benign. The best lives for domesticated animals are on grass, and grass provides the most opportunities for wild animals of all shapes and sizes. In this way, cattle on grass provide habitat for both the domesticated and the wild.

My primary mission these past two decades has been helping, in whatever ways I can, build a more environmentally sound, nourishing, and humane food system. We have a long way to go. I don't urge people to eat meat. But I certainly don't urge refraining from it, either. I encourage omnivorous eaters to seek well-raised meat. Abandoning meat will not positively affect the food system and may diminish one's health. The greatest consumer impact will come from people who eat meat actually *buying it* from good sources. We need to directly support companies, restaurants, farmers, and ranchers who are doing things *the right way*. Nearly everyone in the regenerative food supply chain is operating on a razor-thin margin. In a nation where agribusiness, big food, and pharmaceutical companies hold the political cards, I hold out little hope for any major policy reform. Consumers, however, have enormous power to make positive change in the food system. By buying from good sources, we are voting with our food dollars for the kind of food system we want.

Raising cattle and other grazing animals on grass is inherently resistant to industrialization. This is yet another reason it's an essential part of a more locally and regionally based, more environmentally sustainable food system. Livestock grazing is broadly dispersed, outdoors, and reliant on natural systems. It requires local people, out on the land, with knowledge of climate and ecosystem functions. This is why independent, traditional livestock tenders still exist the world over while croplands have been mostly co-opted by multinational agribusiness corporations growing commodities for the international market.

In the United States, this may be what alarms agribusiness the most about the rising interest in totally grass-fed beef. The huge corporations have gained near-absolute control over the beef markets by buying, shutting down, and consolidating feedlots and slaughterhouses. Although American cattle ranchers remain fiercely independent, they are price takers, not price makers. They control none of the infrastructure that turns cattle into meat and gets it to the end-user. Yet the meat industry does not have, and probably can never have, similar control over cattle raised entirely on grass. This is the province of a growing number of independent farmers and ranchers scattered around the United States, like my husband and me. We raise our own animals from birth and sell beef directly to restaurants, retailers, and consumers.

On another level, I support raising cattle because of the age-old association between domesticated animals and humans beings. Along with tangible nourishment and income, it provides important intangible benefits to humanity. We are better for living alongside them. Those of us who have the pleasure of being around farm animals every day likely benefit the most. We are taught nature's lessons in stark relief. We witness the inevitability of illness, injury, and death; the cycles of birth, growth, aging, and decline. We are constantly reminded of the fragility of life, of what it takes to be a good parent, of bravery, patience, loyalty. If we are paying attention, we are learning from our animals, constantly. Our own impermanence is clearer to us. They also bring unquantifiable yet vast pleasure to many of the people who see them. Year after year, cars driving past our ranch slow down or even stop and get out just to pause and observe our animals. I have been told so often by people in our community how much they enjoy seeing and being near our cattle, turkeys, and goats. "I love walking among your cows," a neighbor told me once at the post office. "They always make me feel so calm."

Humans are strongly drawn to meat and other foods derived from animals. As people gain affluence, to a point, they increase the meat and dairy in their diet. This is happening today in Asia, where demand for meat is growing the fastest. Research also consistently shows about 75 percent of people who give up meat return to it eventually. About 85 percent of people who attempt a strictly vegan diet return to meat. I've met dozens of former vegans or vegetarians in my lifetime. I'm now one myself. People have told me they returned to meat eating for many reasons, often over health issues. For me, it was for many reasons. Mostly, I wanted to make sure I was giving my body everything it needed to remain strong and healthy. Whether for reasons of culture, health, pleasure, or convenience, the overwhelming majority of the world's people eat foods from animals. Clearly, humans will be raising animals for food for the foreseeable future.

Our food system must regenerate itself as nature does. In this pursuit, we should treat animals as indispensable partners, not as inanimate production units. It is time to restore the broken covenant we have with farm animals. Ridding the world of factory farms is an important step. Industrial farming of all types should be replaced with ecological farming systems that recycle nutrients and continually feed and restore earth's living blanket, the soils.

At the heart of this regenerative food system will be cattle and other grazing animals. We will manage their movements, help protect them from predators, make sure they have water and good forage, and attend to their needs. In return, they will help us by keeping our grasslands vibrant— covered in vegetation and teeming with carbon and life belowground. Our cattle will provide us milk and meat, healthful, nutrient-dense foods. We will appreciate and value them for all that they are and all they do.

ACKNOWLEDGMENTS

THIS BOOK HAS BEEN YEARS IN THE MAKING, AND COULD NEVER have happened without the assistance of many colleagues, friends, and family members along the way. Several people have, over the past two decades, generously and patiently shared with me their knowledge of cattle husbandry. To each of you, I want to express my enormous admiration and respect for the honorable way you practice your important profession. In particular, I want to thank: Annie Van Peer, Rob and Michelle Stokes, Ken Bentz Sr., Don McNab, and of course Bill Niman.

Several people kindly took time from their busy schedules to read parts of this manuscript or helped me track down pieces of information. I am indebted to each of you, and am truly grateful for the many valuable suggestions each of you made: Bill Niman, Christine Hahn, Gareth Fisher, and Adam Danforth.

I want to thank my agent, Jennifer Unter, who has unfailingly represented me in the most attentive and professional way; my editor, Ben Watson, who believed in this book from the start, and was responsive, helpful, and professional throughout; and copy editor Laura Jorstad for her careful and thorough review. I wish to thank Stephen Zwick, who has ably and energetically helped spread the messages of this book through his great help with social media.

Most of all I want to thank my wonderful family, Bill, Miles, and Nicholas, each of whom had to get along without me, or with me being distracted, at various times as I toiled away on this book. Bill often did double duty so I would have time to work on the first and second editions of this book. Miles often had to go without the cherished bedtime stories I've read him every night since he was three months old. And Nicholas, more often than not, was nursed in my lap as I researched or edited my writing. I could never have accomplished this task without the patience, love, and support I feel from each of you every day. I am extremely grateful and I love each of you with my whole heart.

NOTES

Introduction

1. R. R. Snapp, *Beef Cattle: Their Feeding and Management in the Corn Belt States*, 3rd ed. (Wiley & Sons, 1939), 16.
2. See N. Hahn Niman, *Righteous Porkchop: Finding a Life and Good Food Beyond Factory Farms* (HarperCollins, 2009), chapter 7, "Beef, the Most (Unfairly) Maligned of Meats."
3. Hahn Niman, *Righteous Porkchop*, chapter 8, "The (Un)-Sacred Milk Cow."
4. The United States is actually fourth (after India, Brazil, and China) in the number of cattle, but it produces the most beef. *BEEF Magazine*, August 13, 2013, beefmagazine .com/cattle-industry-structure/industry-glance-beef-cow-inventory-over-time.

Chapter 1: The Climate Change Case Against Cattle

1. The headline for the press release of *Livestock's Long Shadow* reads: "Rearing Cattle Produces More Greenhouse Gases than Driving Cars," www.un.org/apps/news /story.asp?newsID=20772&CR1=warning#.UpjwIo2f_eI.
2. BBC News, "UN Body to Look at Meat and Climate Link," March 24, 2010, http://news.bbc.co.uk/2/hi/science/nature/8583308.stm.
3. N. Hahn Niman, "The Carnivore's Dilemma," *The New York Times*, October 30, 2009, www.nytimes.com/2009/10/31/opinion/31niman.html?pagewanted=all.
4. US Environmental Protection Agency, Inventory of US Greenhouse Gas Emissions and Sinks: 1990–2017, p. ES-8.
5. Jonathan Sanderman, Tomislav Hengl, and Gregory J. Fiske, "Soil Carbon Debt of 12,000 Years of Human Land Use," *PNAS*, September 5, 2017, www.pnas.org /content/114/36/9575.
6. Adam Majendie and Pratik Parija, "These UN Climate Scientists Think They Can Halt Global Warming for $300 Billion. Here's How," *Time*, October 23, 2019, https://time.com/5709100/halt-climate-change-300-billion.
7. Danielle Prieur, "Could No-Till Farming Reverse Climate Change?," *US News and World Report*, August 4, 2016, www.usnews.com/news/articles/2016-08-04 /could-no-till-farming-reverse-climate-change.
8. "Amazon Besieged: Q&A with Author and Reporter Sue Branford," *Regenetarianism*, June 2, 2019, https://lachefnet.wordpress.com/2019/06/02 /amazon-besieged-an-interview-with-author-and-reporter-sue-branford.
9. Understanding the Soybean Crush, www.cmegroup.com/education/courses /introduction-to-agriculture/grains-oilseeds/understanding-soybean-crush.html.
10. US EPA, Inventory of US Greenhouse Gas Emissions and Sinks: 1990–2017, p. ES-8.
11. J. Schwartz, *Cows Save the Planet* (Chelsea Green Publishing, 2013), 27.

12. "Belching Ruminants, a Minor Player in Atmospheric Methane," Joint FAO/IAEA Programme, http://www-naweb.iaea.org/nafa/news/2008-atmospheric-methane.html.

13. US EPA, Inventory of U.S. Greenhouse Gas Emissions and Sinks: 1990–2017, p. ES-8; and Dr. Howarth's study: https://news.cornell.edu/stories/2019/08/study-fracking-prompts-global-spike-atmospheric-methane.

14. Dr. Myles Allen, speaking at Sustainable Food Trust Conference, Gloucestershire, England, July 5, 2019, www.slideshare.net/Sustainablefoodtrust/myles-allen-154983406.

15. Paul Nicholson, "In Addressing Climate Change, Rice Production Needs More Attention and Urgent Action," Eco-Business, January 9, 2019, www.eco-business.com/opinion/in-addressing-climate-change-rice-production-needs-more-attention-and-urgent-action.

16. "Livestock Manure Management," a report of the US Environmental Protection Agency, www.epa.gov/methane/reports/05-manure.pdf.

17. David Butler, "Frank Mitloehner: Cattle, Climate Change and the Methane Myth," *Alltech*, June 25, 2019, www.alltech.com/features-podcast-blog/frank-mitloehner-cattle-climate-change-and-methane-myth.

18. "Why Methane from Cattle Warms the Climate Differently than CO2 from Fossil Fuels," University of California, Davis, July 7, 2020, https://clear.ucdavis.edu/explainers/why-methane-cattle-warms-climate-differently-co2-fossil-fuels.

19. "Why Methane from Cattle Warms the Climate Differently."

20. "Why Methane from Cattle Warms the Climate Differently."

21. N. Hahn Niman, "Animals Are Essential to Sustainable Food," *Earth Island Journal*, Spring 2010, www.earthisland.org/journal/index.php/eij/article/rancher; cattle numbers are from the 2012 US Department of Agriculture Census of Agriculture (2014), www.agcensus.usda.gov/Publications/2012/#full_report.

22. "Tackling Climate Change Through Livestock," a report of the United Nations Food and Agriculture Organization, September 2013, www.fao.org/3/a-i3437e.pdf.

23. R. A. Leng, *Quantitative Ruminant Nutrition: A Green Science* (1993), www.ciesin.columbia.edu/docs/004-180/004-180.html.

24. Leng, *Quantitative Ruminant Nutrition*.

25. Jennifer Walter, "Feeding Seaweed to Cows Could Curb Their Methane-Laden Burps," *Discover Magazine*, February 10, 2020, www.discovermagazine.com/environment/feeding-seaweed-to-cows-could-curb-their-methane-laden-burps.

26. Geoff Watts, "The Cows That Could Help Fight Climate Change," *BBC Future*, August 6, 2019, www.bbc.com/future/article/20190806-how-vaccines-could-fix-our-problem-with-cow-emissions.

27. Regarding dung beetles, see www.sciencedaily.com/releases/2013/08/130822105031.htm; regarding feeding and genetics for lower methane, see "The Case for Low Methane-Emitting Cattle," *ScienceDaily*, January 10, 2014, www.sciencedaily.com/releases/2014/01/140110131013.htm; regarding weights in the rumen, see "Okine et al. Have Demonstrated a 29% Reduction in Methane Production When Weights Were Added to the Rumen to Stimulate Contraction of the Rumen Wall in Order to Decrease Residence Time of the Feed in the Digestive Tract,"

and E. K. Okine, G. W. Mathison, and R. T. Hardin, "Effects of Changes in Frequency of Reticular Contractions on Fluid and Particulate Passage Rates in Cattle," *Canadian Journal of Animal Science* 67 (1989), 3388; regarding minerals, see "Adequate Mineral Supplementation Is Another Avenue by Which the Cattle Industry May Realize a Net Reduction in Enteric Methane Emissions," and K. Ominski and K. Wittenberg, "Strategies for Reducing Enteric Methane in Forage-Based Beef Production Systems," paper presented at Science of Changing Climates Conference, Edmonton, Alberta, July 2004, 13, www.researchgate.net /publication/255595526_Strategies_for_Reducing_Enteric_Methane_Emissions _in_Forage-Based_Beef_Production_Systems.

28. Bignell et al., eds., *Biology of Termites: A Modern Synthesis* (Springer, 2011).
29. Hizbullah et al., "Methane Emissions from Termites—Landscape Level Estimates and Methods of Measurement," EGU General Assembly, 2003.
30. Horz et al., "Methane-Oxidizing Bacteria in a California Upland Grassland Soil: Diversity and Response to Simulated Global Change," https://aem.asm.org /content/71/5/2642.
31. Zhang, "Moderate Grazing Increases the Abundance of Soil Oxidizing Bacteria," *Journal of Applied Ecology*, May 2019, https://europepmc.org/article/med/31257764.
32. Yong et al., "Impact of Grazing on Shaping Abundance of Active Methanotrophs and Methane Oxidation Activity in a Grassland Soil," *Biology and Fertility of Soils*, April 2020.
33. Website of the US EPA, http://epa.gov/climatechange/ghgemissions/gases/n2o.html.
34. US Energy Information Administration, "Emissions of Greenhouse Gases in the United States," https://www.eia.gov/environment/emissions/ghg_report/ ghg_nitrous.php.
35. US EIA, "Emissions of Greenhouse Gases in the United States."
36. Robert Sanders, "Fertilizer Use Responsible for Increase in Nitrous Oxide in Atmosphere," news release of the UC Berkeley News Center, April 2, 2012, https://newscenter.berkeley.edu/2012/04/02/fertilizer-use-responsible-for -increase-in-nitrous-oxide-in-atmosphere.
37. David Kanter et al., *Drawing Down N₂O to Protect Climate and the Ozone Layer*, a United Nations Environment Program report, November 2013, 21, www. researchgate.net/publication/268213461_Drawing_Down_N2O_to_Protect _Climate_and_the_Ozone_Layer_A_UNEP_Synthesis_Report.
38. S. Fairlie, *Meat: A Benign Extravagance* (Chelsea Green Publishing, 2011), 159.
39. Website of the US EPA, http://epa.gov/climatechange/ghgemissions/sources /agriculture.html.
40. Fairlie, *Meat*, 161.
41. Fairlie, *Meat*, 161.
42. *The Emissions Gap Report, 2013*, a United Nations Environment Program report, November 2013, xvi, www.unep.org/resources/emissions-gap-report-2013.
43. "Livestock and Climate Change," *World Watch* (November–December 2009), https://awellfedworld.org/wp-content/uploads/Livestock-Climate-Change -Anhang-Goodland.pdf.

44. From an interview with Robert Goodland, http://juliansstory.wordpress.com /2011/10/27/meeting-robert-goodland.

45. "ERS Report: Interdependence of China, United States, and Brazil in Soybean Trade," Illinois Farm Policy News, June 30, 2019.

46. "How Much US Meat Comes from Foreign Sources," a report of the Economic Research Service of the US Department of Agriculture, September 20, 2012, www.ers .usda.gov/amber-waves/2012/september/how-much-us-meat.

47. Hahn Niman, *Righteous Porkchop*, 136–40.

48. H. Biswell, *Prescribed Burning in California Wildlands Vegetation Management* (University of California Press, 1989).

49. Biswell, *Prescribed Burning in California Wildlands Vegetation Management*, 48.

50. Biswell, *Prescribed Burning in California Wildlands Vegetation Management*, 49–50.

51. C. Mann, *1491* (Vintage, 2006), 285–86.

52. D. Montgomery, *Dirt: The Erosion of Civilizations* (University of California Press, 2012), 29.

53. See Montgomery, *Dirt*.

54. *US Forest Facts and Historical Trends*, a report of the Forest Service of the US Department of Agriculture, September 2001, www.fia.fs.fed.us/library/briefings -summaries-overviews/docs/ForestFactsMetric.pdf.

55. "Deforestation," *Encyclopedia Britannica*, www.britannica.com/EBchecked/topic /155854/deforestation/306437/Effects.

56. L. Palmer, "In the Pastures of Colombia, Cows, Crops and Timber Coexist," *Yale Environment 360*, March 13, 2014, http://e360.yale.edu/feature/in_the_pastures _of_colombia_cows_crops_and_timber_coexist/2746.

57. Lela Nargi, "Silvopasture Can Mitigate Climate Change," *Civil Eats*, January 7, 2019, https://civileats.com/2019/01/07/silvopasture-can-mitigate-climate-change -will-u-s-farmers-take-it-seriously.

58. Nargi, "Silvopasture Can Mitigate Climate Change."

59. Palmer, "In the Pastures of Colombia, Cows, Crops and Timber Coexist."

60. See, for instance, S. Spiegal, L. Huntsinger, P. Hopkinson, J. Bartolome, and S. Sheri, "Overview of California Range Ecosystems," in *Ecosystems of California*, ed. H. Mooney and E. Zavaleta (University of California Press, 2016).

61. G. Azeez, "Soil Carbon and Organic Farming: Summary of Findings," a report of the U.K. Soil Association, 2009, www.soilassociation.org/LinkClick.aspx ?fileticket=BVTfaXnaQYc%3D&.

62. F. M. Mitloehner, "Clearing the Air on 'Livestock and Climate Change,'" 2010, www.ansci.cornell.edu/cnconf/2010proceedings/CNC2010.5.Mitloehner.pdf.

63. "The Unusual Uses for Animal Body Parts," *BBC News*, June 7, 2011.

64. Schwartz, *Cows Save the Planet*, 12.

65. "Tackling Climate Change Through Livestock," FAO, 41.

66. "Tackling Climate Change Through Livestock," FAO, xiii (emphasis added).

67. See, for instance, A. Gattinger et al., "Enhanced Top Soil Carbon Stocks Under Organic Farming," *Proceedings of the National Academy of Sciences*, October 2012, www.pnas.org/content/109/44/18226.full.

68. Cited in Azeez, "Soil Carbon and Organic Farming": "An estimated 89% of the global potential for agricultural greenhouse gas mitigation would be through carbon sequestration." P. Smith et al., "Greenhouse Gas Mitigation in Agriculture," *Philosophical Transactions of the Royal Society of London, Series B Biological Sciences*, 2008.
69. Azeez, "Soil Carbon and Organic Farming."
70. Montgomery, *Dirt*, 23.
71. Montgomery, *Dirt*, 3.
72. J. Schwartz, "Soil as Carbon Storehouse: New Weapon in Climate Fight?" *Yale Environment 360*, March 4, 2014, http://e360.yale.edu/feature/soil_as _carbon_storehouse_new_weapon_in_climate_fight/2744.
73. Azeez, "Soil Carbon and Organic Farming."
74. Cornelia Rumpel et al., "'4per1,000' Initiative will Boost Soil Carbon for Climate and Food Security," *Nature*, January 2018, www.nature.com/articles/d41586 -017-09010-w.
75. A. Gillmullina et al., "Management of Grasslands by Mowing Versus Grazing," *Applied Science Ecology*, June 2020.
76. "California's Latest Weapon Against Climate Change Is Low-Tech Farm Soil," National Public Radio, May 2, 2019, www.npr.org/2019/05/02/718736830 /californias-latest-weapon-against-climate-change-is-low-tech-farm-soil.
77. Bronson Griscom et al., "Natural Climate Solutions," *PNAS*, October 31, 2017, www.pnas.org/content/114/44/11645.
78. Griscom et al., "Natural Climate Solutions." See also D. Pimentel and M. Pimentel, "[A] Feedback Mechanism May Exist Wherein Increased Global Warming Intensifies Rainfall, Which, in Turn, Increases Erosion and Continues the Cycle," *Food, Energy, and Society*, 3rd ed. (CRC Press, 2008), 212, citing Lal, 2002.
79. Griscom et al., "Natural Climate Solutions."
80. "Glomalin: Hiding Place for a Third of the World's Stored Soil Carbon," *Agricultural Research*, September 2002, www.ars.usda.gov/is/ar/archive/sep02 /soil0902.htm; "Glomalin: What Is It . . . and What Does It Do?" *Agricultural Research*, July 2008, www.ars.usda.gov/is/AR/archive/jul08/glomalin0708.pdf.
81. See, for instance, S. Fonte et al., "Earthworm Populations in Relation to Soil Organic Matter Dynamics and Management in California Tomato Cropping Systems," *Applied Soil Ecology*, 2009, http://ucanr.org/sites/ct/files/44375.pdf.
82. Azeez, "Soil Carbon and Organic Farming."
83. USDA *Agricultural Research* articles.
84. Personal communication with author, April 22, 2014.
85. Azeez, "Soil Carbon and Organic Farming."
86. Azeez, "Soil Carbon and Organic Farming."
87. L. Manske, "Effects of Grazing Management Treatments on Rangeland Vegetation," North Dakota State University, website of the Dickinson Research Extension Center, 2004, www.ag.ndsu.edu/archive/dickinso/research/2003/range03c.htm.
88. D. Harnett, paper presented to the 1999 Society for Range Management Meeting, February 1999, website of the Ecological Society of America, www.esa.org/science _resources/publications/purePrairieLeague.php.

89. Hahn Niman, "Animals Are Essential."

90. R. Lal and B. A. Stewart, *Food Security and Soil Quality* (CRC Press, 2010), 255, www.academia.edu/5208901/Food_Security_and_Soil_Quality_By_Rattan _Lal_B_A._Stewart.

91. Lal and Stewart, *Food Security and Soil Quality*, 263.

92. Lal and Stewart, *Food Security and Soil Quality*, 251.

93. L. B. Guo and R. M. Gifford, "Soil Carbon Stocks and Land Use Change: A Meta Analysis," *Global Change Biology*, November 23, 2002, http://onlinelibrary .wiley.com/doi/10.1046/j.1354-1013.2002.00486.x/abstract.

94. Azeez, "Soil Carbon and Organic Farming."

95. Schwartz, *Cows Save the Planet*, 15; and Schwartz, "Soil as Carbon Storehouse."

96. Schwartz, *Cows Save the Planet*, 15.

97. "Creating Topsoil by Dr. Christine Jones," March 24, 2006, http://creatingnewsoil .blogspot.com.

98. A. Savory, "Reversing Global Warming While Meeting Human Needs: An Urgently Needed Land-Based Option," speech given at Tufts University, January 25, 2013.

99. Savory, "Reversing Global Warming While Meeting Human Needs."

100. Savory, "Reversing Global Warming While Meeting Human Needs."

101. "The Savory Institute: Healing the World's Grasslands, Rangelands, and Savannas," Allan Savory interview with *World Watch*, April 15, 2011, http://blogs.worldwatch .org/nourishingtheplanet/tag/international-society-for-range-management.

102. Savory, "Reversing Global Warming While Meeting Human Needs."

103. Savory, "Reversing Global Warming While Meeting Human Needs."

104. J. Marty, "Effects of Cattle Grazing on Diversity in Ephemeral Wetlands," *Conservation Biology* 19, no. 5 (October 2005), citing White, 1979; Sousa, 1984; Hobbs and Huenneke, 1992, www.elkhornsloughctp.org/uploads /files/1401307451Marty.%202005.%20Effects%20of%20cattle%20grazing%20 on%20diversity%20in%20ephemeral%20wetlands..pdf.

105. Savory, *World Watch* interview.

106. Savory, *Holistic Resource Management* (Island Press, 1988), 38.

107. Savory, *Holistic Resource Management*, 35, 38.

108. Savory, *Holistic Resource Management*, 39–40.

109. Savory, *Holistic Resource Management*, 37.

110. Savory, *Holistic Resource Management*, 37.

111. Savory, *Holistic Resource Management*, 38.

112. Savory, *Holistic Resource Management*, 37.

113. A. Savory, talk given at the Restoring Working Landscapes, Producing Sustainable Meat conference, TomKat Ranch, Pescadero, California, December 3, 2013.

114. Savory, "Reversing Global Warming While Meeting Human Needs."

115. Schwartz, "Soil as Carbon Storehouse."

116. S. Itzkan, "Hut with a View," Seth Itzkan Reports from the Africa Centre for Holistic Management in Zimbabwe, http://hutwithaview.com.

117. Savory, *Holistic Resource Management*, 9.

118. Savory, TomKat Ranch.

119. Savory, *Holistic Resource Management*, 34–35.

120. Savory, "Reversing Global Warming While Meeting Human Needs."

121. Savory, *World Watch* interview.

122. "Animals Key to Biodiversity of Over-Fertilized Prairies," University of Maryland press release, March 10, 2014, www.umdrightnow.umd.edu/news/animals-key-to-biodiversity-over-fertilized-prairies.

123. Savory, TomKat Ranch.

124. Savory, "Reversing Global Warming While Meeting Human Needs."

125. Savory, *World Watch* interview.

126. Savory, TomKat Ranch.

127. Savory, *World Watch* interview.

128. Savory, TomKat Ranch.

129. Savory, TomKat Ranch.

130. Savory, "Reversing Global Warming While Meeting Human Needs."

131. Teague et al., "Grazing Management Impacts on Vegetation, Soil Biota and Soil Chemical, Physical and Hydrological Properties in Tall Grass Prairie," *Agriculture, Ecosystems and Environment*, 2011, www.sciencedirect.com/science/article/pii/S0167880911000934.

132. Weber et al., "Effect of Gracing on Soil-Water Content in Semiarid Rangelands of Southeast Idaho," *Journal of Arid Environments* 141 (May 2011), 310, www.sciencedirect.com/science/article/pii/S0140196310003460.

133. S. Itzkan, "Upside (Drawdown): The Potential of Restorative Grazing to Mitigate Global Warming by Increasing Carbon Capture on Grasslands," a report of PlanetTech Associates, April 2014, www.planet-tech.com/upsidedrawdown#sthash.mKkNI9fj.dpuf.

134. "Study: White Oak Pastures Beef Reduces Atmospheric Carbon," White Oak Pastures, June 4, 2019, http://blog.whiteoakpastures.com/blog/carbon-negative-grassfed-beef.

135. W. R. Teague et al., "The Role of Ruminants in Reducing Agriculture's Carbon Footprint in North America," *Journal of Soil and Water Conservation* 71, no. 2 (2016), 156–64, http://www.jswconline.org/content/71/2/156.full.pdf+html.

136. Itzkan, "Upside (Drawdown)."

137. E. Coleman, "Debunking the Meat/Climate Change Myth," August 7, 2009, http://grist.org/article/2009-08-07-debunking-meat-climate-change-myth.

138. "Tackling Climate Change," FAO.

139. "Tackling Climate Change," FAO, 23.

140. "Tackling Climate Change," FAO, 17.

141. M. Lee, Sustainable Food Trust Conference, Gloucestershire, England, July 5, 2019.

142. Lal and Stewart, *Food Security and Soil Quality*, 260.

143. Lal and Stewart, *Food Security and Soil Quality*, 184.

144. R. Tuhus-Dubrow, "How to Solve Climate Change with Cows (Maybe)," *Boston Globe*, May 4, 2014, www.bostonglobe.com/ideas/2014/05/03/how-solve-climate-change-with-cows-maybe/j3c4uoHv4iJqWjHXezlonN/story.html.

145. "Tackling Climate Change," FAO, 17.
146. E. Rosenthal, "To Cut Global Warming, Swedes Study Their Plates," *New York Times*, October 22, 2009, www.nytimes.com/2009/10/23/science/earth/23degrees.html.
147. J. Hendrickson, *Energy Use in the US Food System: A Summary of Existing Research and Analysis*, a report of the Center for Integrated Agricultural Systems, University of Wisconsin–Madison, July 2008, www.cias.wisc.edu/wp-content /uploads/2008/07/energyuse.pdf.
148. *Organic Works: Providing More Jobs Through Organic Farming and Local Food Supply*, a report of the UK Soil Association (Bristol House, 2006).
149. Note that "Tackling Climate Change," FAO, 17, states: "Emissions associated with energy consumption (directly or indirectly related to fossil fuel) are mostly related to feed production, and fertilizer manufacturing, in particular. When added up along the chains, energy use contributes about 20 percent of total sector emissions."
150. S. Murray, "The Deep Fried Truth," *New York Times*, December 14, 2007, www.nytimes.com/2007/12/14/opinion/14murray.html.
151. N. Hahn Niman and B. Niman, "The Cost of Wasted Food," *The Atlantic*, December 2, 2009, www.theatlantic.com/health/archive/2009/12/the-cost-of -wasted-food/31089.
152. "World War II Rationing," *United States History*, www.u-s-history.com/pages /h1674.html.
153. See interview with William Brangham, *PBS Newshour*, March 1, 2019.
154. "How Much Do Our Wardrobes Cost the Environment," *The World Bank*, September 23, 2019, www.worldbank.org/en/news/feature/2019/09/23/costo -moda-medio-ambiente.
155. Said to Margot Einstein after his sister's Maja's death, 1951; quote by Hanna Loewy in A&E Television, Einstein Biography, VPI International, 1991.

Chapter 2: All Food Is Grass

1. For the 40 percent figure, see World Resources Institute, www.wri.org/publica-tion/content/8269. For the 70 percent figure, see: "Are Grasslands Under Threat?," an analysis of the United Nations Food and Agriculture Organization, www.fao .org/uploads/media/grass_stats_1.pdf.
2. Stromberg et al., eds., *California Grasslands: Ecology and Management* (University of California Press, 2007), 7.
3. Pimentel and Pimentel, "[A] Feedback Mechanism May Exist," 363.
4. Montgomery, *Dirt*, x.
5. Montgomery, *Dirt*, xiv.
6. Montgomery, *Dirt*, 16.
7. Montgomery, *Dirt*, 15.
8. J. Rogers and P. G. Feiss, *People and the Earth: Basic Issues in the Sustainability of Resources and Environment* (Cambridge University Press, 1998), 63. See also: "About one-third of the topsoil from US agricultural land already has been lost," Pimentel and Pimentel, "[A] Feedback Mechanism May Exist," 41.

9. J. P. Curry, *Grassland Invertebrates: Ecology, Influence on Soil Fertility and Effects on Plant Growth* (Springer, 1993).

10. Pimentel and Pimentel, "[A] Feedback Mechanism May Exist," 190.

11. Montgomery, *Dirt*, 17. See also: "Along with plants and animals, microbes are a vital component of the soil and constitute a large percentage of the soil biomass. One square meter of soil may support about 200,000 arthropods and enchytraeids, plus billions of microbes (Wood, 1989; Lee and Foster, 1991) . . . In addition, soil bacteria and fungi add 4000–5000 species and in this way contribute significantly to the biodiversity . . ." Pimentel and Pimentel, "[A] Feedback Mechanism May Exist," 209.

12. F. Troeh et al., *Soil and Water Conservation*, 3rd ed. (Prentice-Hall, 1999), 332. See also: "Vegetative cover is the principal way to protect soil and water resources," Pimentel and Pimentel, "[A] Feedback Mechanism May Exist," 31.

13. Troeh, *Soil and Water Conservation*.

14. Lal and Stewart, *Food Security and Soil Quality*, 152. See also: "[E]roded soil absorbs 87% less water by infiltration than uneroded soils." Pimentel and Pimentel, "[A] Feedback Mechanism May Exist," 190, citing Guenette, 2001.

15. Pimentel and Pimentel, "[A] Feedback Mechanism May Exist," 203.

16. W. Berry, *New Roots of Agriculture* (University of Nebraska Press, 1980), xii.

17. Pimentel and Pimentel, "[A] Feedback Mechanism May Exist," 31.

18. Berry, xiii.

19. Pimentel and Pimentel, "[A] Feedback Mechanism May Exist," 205.

20. Pimentel and Pimentel, "[A] Feedback Mechanism May Exist."

21. Pimentel and Pimentel, "[A] Feedback Mechanism May Exist," 206.

22. G. Boody et al., "Multifunctional Agriculture in the United States," *Bioscience* 55 (2005), 32.

23. Lal and Stewart, *Food Security and Soil Quality*, 153 (internal citation omitted).

24. See "Fundamentals of Cation Exchange Capacity," https://www.extension.purdue.edu/extmedia/ay/ay-238.html.

25. Lal and Stewart, *Food Security and Soil Quality*, 252.

26. Genesis 3:19.

27. Psalms 104.

28. Deuteronomy 11:15.

29. J. J. Ingalls, "In Praise of Bluegrass" (1872), http://grassbydesign.com/pdf/douglas.pdf.

30. Stromberg et al., *California Grasslands*, 37.

31. "Last Time Carbon Dioxide Levels Were This High: 15 Million Years Ago," *Science Daily*, October 9, 2009, www.sciencedaily.com/releases/2009/10/091008152242.htm.

32. Stromberg et al., *California Grasslands*, 37.

33. Stromberg et al., *California Grasslands*, 50–52.

34. Stromberg et al., *California Grasslands*, 49–52.

35. M. Kinver, "Ecologists Learn Lessons from the Ghosts of Megafauna," *BBC News*, March 24, 2014, www.bbc.com/news/science-environment-26718199.

36. "Big Game Could Roam US Plains," *BBC News*, August 18, 2005, http://news.bbc.co.uk/2/hi/science/nature/4160560.stm.

37. S. Daley, "From Untended Farmland, Reserve Tries to Recreate Wilderness from Long Ago," *New York Times*, June 13, 2014, www.nytimes.com/2014/06/14/world /europe/from-untended-farmland-reserve-tries-to-recreate-wilderness-from-long -ago.html?emc=edit_th_20140614&nl=todaysheadlines&nlid=28644965&_r=0.

38. Daley, "From Untended Farmland."

39. Stromberg et al., *California Grasslands*, 2.

40. See Stromberg et al., *California Grasslands*, 57, 58, 59, 64, 65, 75, 175, 209, 219, 256.

41. Hahn Niman, *Righteous Porkchop*, 136, citing regarding cattle zoological classification, C. Hickman, *Integrated Principles of Zoology*, 7th ed. (Times Mirror / Mosby College Publishing, 1994), 662; regarding domestication, P. Davis and A. Dent, *Animals That Changed the World: The Story of Domestication of Wild Animals* (Crowell-Collier Press, 1968), 66, and A. Beja-Pereria et al., "The Origin of European Cattle: Evidence from Modern and Ancient DNA," *Proceedings of the National Academy of Sciences*, May 11, 2006.

42. Snapp, *Beef Cattle*, 5, citing *Yearbook of Agriculture 1921*, US Department of Agriculture (US Government Printing Office, 1921), 232.

43. Snapp, *Beef Cattle*, 16.

44. Montgomery, *Dirt*, 18.

45. H. D. Hughes, *Forages* (Iowa State College Press, 1951), 22.

46. C. P. Anderson, *Grass: Yearbook of Agriculture 1948*, US Department of Agriculture (US Government Printing Office, 1948), v.

47. Hughes, *Forages*, 8.

48. Montgomery, *Dirt*, 173; Pimentel and Pimentel, "[A] Feedback Mechanism May Exist," 41.

49. D. Pimentel and N. Kounang, "Ecology of Soil Erosion in Ecosystems," *Ecosystems* 1 (1998), 416.

50. "Losing Ground," a report of the Environmental Working Group, April 2011, http://static.ewg.org/reports/2010/losingground/pdf/losingground_report.pdf.

51. D. Barber, "What Farm-to-Table Got Wrong," *New York Times*, May 17, 2014.

52. See, for instance, N. Hahn Niman, "Biofuels: Bad News for Animals," *AWI Quarterly*, Summer 2008, https://awionline.org/content-types-orchid-legacy/awi -quarterly/biofuels-bad-news-animals.

53. J. Zuckerman, "Plowed Under," *American Prospect*, May 2014, http://prospect.org /article/plowed-under.

54. L. Kreiger, "California Drought: San Joaquin Valley Sinking as Farmers Race to Tap Aquifer," *San Jose Mercury News*, March 29, 2014, www.mercurynews.com /drought/ci_25447586/california-drought-san-joaquin-valley-sinking-farmers-race.

55. See F. Thicke, *A New Vision for Iowa Food and Agriculture* (Mulberry Knoll Books, 2010).

56. Lal and Stewart, *Food Security and Soil Quality*, 130.

57. J. Lundgren, Keynote Address, EcoFarm Conference, January 22, 2020.

58. Patrick Holden, founder of the non-profit organization The Sustainable Food Trust, has spent years advancing the idea of True Cost Accounting. He urges that policymakers at all levels must take into consideration all effects their decisions

have related to food and farming. In his blog, Holden as seeking "to assess the costs and benefits of different food production systems. In doing so, those using production methods that are detrimental to the environment and society would have to pay for the damage they do, while those that are sustainable and deliver a wide range of benefits would be rewarded. This should ultimately have the effect of making food produced in a damaging way expensive, or ideally phased out, whilst sustainable food could become more affordable." He has worked with the United Nations and elsewhere to advocate for use of this method. To learn more, see https://sustainablefoodtrust.org/articles/true-cost-food/.

Chapter 3: Water

1. C. Dell'Amore, "Biggest Dead Zone Ever Forecast in Gulf of Mexico: Oxygen-Deprived Area May Be Size of New Jersey, Scientists Say," *National Geographic News*, June 24, 2013, http://news.nationalgeographic.com/news /2013/06/130621-dead-zone-biggest-gulf-of-mexico-science-environment.
2. See, for instance, Hahn Niman, *Righteous Porkchop*, 49–51.
3. See T. Charles and H. Stuart, *Commercial Poultry Farming* (Interstate Printing, 1936), v.
4. Federal Water Pollution Control Act Amendments of 1971, Supplemental Views of Senator Robert Dole, Legal Compilation, Environmental Protection Agency, January 1973.
5. Pimentel and Pimentel, "[A] Feedback Mechanism May Exist," 232.
6. Pimentel and Pimentel, "[A] Feedback Mechanism May Exist," 210.
7. W. Jackson, *New Roots for Agriculture* (University of Nebraska Press, 1980), 20.
8. See J. Burkholder, "Impacts of Waste from Concentrated Animal Feeding Operations on Water Quality," *Environmental Health Perspectives*, February 2007, www.ncbi.nlm.nih.gov/pmc/articles/PMC1817674.
9. "Concentrated Animal Feeding Operations: Health Risks from Water Pollution," a publication of the Institute for Agriculture and Trade Policy (IATP), www.iatp .org/files/421_2_37390.pdf.
10. Today the vast majority (over 90 percent) of American livestock and poultry feed is corn. You often see the figure that 70 percent of US grain is fed to our livestock and poultry, but this figure is out of date. About 40 percent of corn grown in the United States is now being used for ethanol. A large portion is also exported. See J. Foley, "It's Time to Rethink America's Corn System." *Scientific American*, March 5, 2013, www.scientificamerican.com/article/time-to-rethink-corn, and L. Hoffman et al. "Feed Grains Backgrounder," a report of the Economic Research Service of the US Department of Agriculture, March 2007, www.ers.usda.gov /media/197317/fds07c01_1_.pdf.
11. S. Thompson, "Running on Empty?: 'Great Ethanol Debate' Waged at NCGA Forum," *Rural Cooperatives Magazine*, a publication of the US Department of Agriculture, September–October 2005, http://www.rurdev.usda.gov/rbs/pub /sep05/running.htm.
12. Pimentel and Pimentel, "[A] Feedback Mechanism May Exist," 317.

13. "Fertilizer Use and Price," a report of the US Department of Agriculture, Economic Research Service, July 12, 2013, www.ers.usda.gov/data-products /fertilizer-use-and-price.aspx#.U3AfAy_c0Xw.

14. "Putting Dairy Cows Out to Pasture: An Environmental Plus," a report of the US Department of Agriculture, Agricultural Research Service, April 29, 2011, www .ars.usda.gov/is/ar/2011/may11/cows0511.htm.

15. Hendrickson, *Energy Use in the US Food System*; "Agriculture's Supply and Demand for Energy and Energy Products," a report by the US Department of Agriculture, Economic Research Service, May 2013, www.ers.usda.gov/ media/1104145/eib112.pdf.

16. Hendrickson, *Energy Use in the US Food System*; "Agriculture's Supply and Demand for Energy and Energy Products."

17. Hendrickson, *Energy Use in the US Food System*; "Agriculture's Supply and Demand for Energy and Energy Products."

18. Pimentel and Pimentel, "[A] Feedback Mechanism May Exist," 153.

19. Pimentel and Pimentel, "[A] Feedback Mechanism May Exist," 161.

20. "Pesticides in the Nation's Streams and Groundwater," a fact sheet of the US Geological Survey, March 2006, http://pubs.usgs.gov/fs/2006/3028 (emphasis added).

21. "Pesticides in the Nation's Streams and Groundwater."

22. "Pesticides in the Nation's Streams and Groundwater."

23. K. Nichols, "The Role of Soil Biology in Improving Soil Quality," a webinar of National Resources Conservation Service / US Department of Agriculture, September 13, 2012, https://www.youtube.com/watch?v=eGxjcxVMbsg.

24. "Concentrated Animal Feeding Operations," IATP.

25. See, for instance, the Pew Charitable Trusts' Campaign on Human Health and Industrial Farming: www.pewtrusts.org/en/projects/campaign-on-human -health-and-industrial-farming.

26. C. Dean, "Drugs Are in the Water. Does It Matter?" *New York Times*, April 3, 2007.

27. See Hahn Niman, *Righteous Porkchop*, chapter 8.

28. J. R. Winsten et al., "Differentiated Dairy Grazing Intensity in the Northeast," *Journal of Dairy Science* 83 (2000), 836.

29. Winsten et al., "Differentiated Dairy Grazing Intensity."

30. M. Halverson, *Farm Animal Health and Well-Being*, a report prepared for the Minnesota Planning Agency, Environmental Quality Board, April 23, 2001, 120.

31. Von Keyserlingk et al., "The Welfare of Dairy Cattle," *Journal of Dairy Science*, September 2009, https://www.sciencedirect.com/science/article/pii/S002203 0209707350; see also N. Hahn Niman, "The Unkindest Cut," *New York Times*, March 7, 2005, https://www.nytimes.com/2005/03/07/opinion/the-unkindest -cut.html.

32. R. Marks, *Cesspools of Shame*, a report by the Natural Resources Defense Council, July 2001, 14, www.nrdc.org/water/pollution/cesspools/cesspools.pdf.

33. R. Whitlock, *A Short History of Farming in Britain* (John Baker Ltd., 1965).

34. Snapp, *Beef Cattle*, 16.
35. Data from the 2012 US Department of Agriculture Census of Agriculture, May 2014, www.agcensus.usda.gov/Publications/2012.
36. Mooney and Zavaleta, *Ecosystems of California*, 37 (internal citation omitted).
37. "Sustainable Farming Systems: Demonstrating Environmental and Economic Performance," a study of the University of Minnesota et al., June 2001.
38. "How Much Water to Make One Pound of Beef," https://www.vegsource.com /articles/pimentel_water.htm.
39. "How Much Water to Make One Pound of Beef," https://www.vegsource.com /articles/pimentel_water.htm.
40. R. Smithers, "Food Products Should Carry 'Water Footprint' Information, Says Report," *Guardian*, July 20, 2009, www.guardian.co.uk/environment/2009/jul/20 /food-water-footprint.
41. Beckett and Oltjen, "Estimation of the Water Requirement for Beef Production in the United States," *Journal of Animal Science* 71 (1993), 818.
42. L. Keith, *The Vegetarian Myth* (PM Press, 2009), 102.
43. L. Keith, *The Vegetarian Myth* (PM Press, 2009), 102. See also: "[A]nimal protein has about 1.4 times the biological value as food compared with grain protein," Pimentel and Pimentel, "[A] Feedback Mechanism May Exist," 70.
44. Boody, "Multifunctional Agriculture in the United States," 32.
45. "Managing Runoff and Erosion on Croplands and Pastures," a publication of the University of Georgia Cooperative Extension, March 2009, www.caes.uga.edu /applications/publications/files/pdf/B%201152-15_2.PDF.

Chapter 4: Biodiversity

1. H. Steinfeld et al., *Livestock's Long Shadow: Environmental Issues and Options*, a report of the United Nations Food and Agriculture Organization, 2007, 254.
2. Schwartz, *Cows Save the Planet*, 193.
3. P. Krausman, "An Assessment of Rangeland Activities on Wildlife Populations and Habitats," a publication of the National Resource Conservation Service, 2011, 257, www.nrcs.usda.gov/Internet/FSE_DOCUMENTS/stelprdb1045801.pdf (internal citations omitted).
4. Krausman, "An Assessment of Rangeland Activities," 259.
5. Krausman, "An Assessment of Rangeland Activities," citing Licht, 1997; Higgins et al., 2002.
6. Marty, "Effects of Cattle Grazing on Diversity," citing Noy-Meir et al., 1989; Harrison, 1999; Hayes and Holl, 2003, http://vernalpools.net/documents/Marty %20Cons%20Bio.pdf.
7. Marty, "Effects of Cattle Grazing on Diversity," citing McNaughton et al., 1989; Milchunas and Lauenroth, 1993; Perevolotsky and Seligman, 1998.
8. Marty, "Effects of Cattle Grazing on Diversity," citing Collins et al., 1998; Harrison, 1999; Maestas et al., 2003.
9. "Animals Key to Biodiversity of Over-Fertilized Prairies," March 10, 2014, www .umdrightnow.umd.edu/news/animals-key-biodiversity-over-fertilized-prairies;

see also "How Grazing Helps Plant Diversity," 2014, www.sciencedaily.com/
releases/2014/03/140309150536.

10. R. Hart, "Plant Biodiversity on Shortgrass Steppe After 55 Years of Zero, Light,
Moderate, or Heavy Cattle Grazing," *Plant Ecology*, July 2001, http://link.springer.
com/article/10.1023/A:1013273400543.

11. "Vegetation Change After 65 Years of Grazing and Grazing Exclusion," *Journal of
Rangeland Management*, November 2004, www.cabnr.unr.edu/news/story
.aspx?StoryID=295.

12. "Vegetation Change After 65 Years of Grazing and Grazing Exclusion."

13. Mooney and Zavaleta, *Ecosystems of California*, 23.

14. "Vernal Pools: Liquid Sapphires of the Chaparral," website of the California
Chaparral Institute, https://www.californiachaparral.org/chaparral/vernal-pools/.

15. Marty, "Effects of Cattle Grazing on Diversity."

16. The Nature Conservancy newsletter, Fall 2009, www.nature.org/ourinitiatives
/regions/northamerica/unitedstates/california/fall-2009-newsletter-1-1.pdf.

17. Krausman, "An Assessment of Rangeland Activities," 262.

18. Marty, "Effects of Cattle Grazing on Diversity" (internal citations omitted).

19. "Grazing CRP Land Improves Feed, Habitat," a case study of the Leopold
Center for Sustainable Agriculture, www.leopold.iastate.edu/sites/default/files
/GrazingCRPcasestudy.pdf.

20. Hahn Niman, "Biofuels: Bad News for Animals."

21. Krausman, "An Assessment of Rangeland Activities," 262, citing Kohler and
Rauer, 1991.

22. "What's Good for the Herd Is Good for the Bird," *BEEF Magazine*, December
18, 2019.

23. Mooney and Zavaleta, *Ecosystems of California*, 30–31.

24. Mooney and Zavaleta, *Ecosystems of California*, 30–31.

25. US Fish and Wildlife Service data. www.fws.gov/pollinators.

26. Data taken from a report by biologists at UC Berkeley. See my article on the study:
N. Hahn Niman, "A Way to Save America's Bees: Buy Free Range Beef," *The
Atlantic*, July 14, 2011, www.theatlantic.com/health/archive/2011/07/a-way-to-save
-americas-bees-buy-free-range-beef/241935.

27. M. Wines, "Soaring Bee Deaths in 2012 Sound Alarm on Malady," *New York
Times*, March 28, 2013, www.nytimes.com/2013/03/29/science/earth/soaring-
bee-deaths-in-2012-sound-alarm-on-malady.html.

28. Hahn Niman, "A Way to Save America's Bees."

29. Hahn Niman, "A Way to Save America's Bees."

30. Krausman, "An Assessment of Rangeland Activities," 279.

31. C. Mendenhall et al., "Predicting Biodiversity Change and Averting Collapse in
Agricultural Landscapes," *Nature*, May 8, 2014, www.nature.com/nature/journal
/vaop/ncurrent/full/nature13139.html.

Chapter 5: Overgrazing

1. Montgomery, *Dirt*, 236.

2. Montgomery, *Dirt*, xii.

3. R. Lal, "Potential of Desertification Control to Sequester Carbon and Mitigate the Greenhouse Effect," *Climatic Change* 51, no. 1 (October 2001), 35–72.

4. S. Rosen, "Desertification and Pastoralism: A Historical Review of Pastoral Nomadism in the Negev Region," Encyclopedia of Life Support Systems.

5. Montgomery, *Dirt*, xx.

6. Montgomery, *Dirt*, xii.

7. Montgomery, *Dirt*, 21.

8. Montgomery, *Dirt*, 4.

9. Jackson, *New Roots for Agriculture*, 2.

10. Jackson, *New Roots for Agriculture*.

11. Jackson, *New Roots for Agriculture*, 12.

12. Montgomery, *Dirt*, 5.

13. Montgomery, *Dirt*, 23.

14. Montgomery, *Dirt*, 23–24.

15. Jackson, *New Roots for Agriculture*, 12–13.

16. P. Starrs, *Let the Cowboy Ride* (Johns Hopkins University Press, 1998), 21.

17. Starrs, *Let the Cowboy Ride*.

18. "Fact Sheet on the BLM's Management of Livestock Grazing," a document of the US Bureau of Land Management, www.blm.gov/wo/st/en/prog/grazing.html.

19. Decision Notice and Finding of No Significant Environmental Impact, Forest Service of the US Department of Agriculture, September 2010, 59, www.fs.usda.gov/Internet/FSE_DOCUMENTS/stelprdb5200869.pdf (internal citations omitted).

20. R. Mearns, "Livestock and Environment: Potential for Complementarity," www.fao.org/docrep/w5256t/w5256t02.htm.

21. K. Weber and S. Horst, "Desertification and Livestock Grazing: The Roles of Sedentarization, Mobility and Rest," *Pastoralism*, October 2011, citing Seligman and Perevolotsky, 1994; Olaizola et al., 1999; Cummins, 2009. www.pastoralism journal.com/content/1/1/19#.

22. See Lal and Stewart, *Food Security and Soil Quality*.

23. Montgomery, *Dirt*, 208.

24. L. Sorensen, "Colorado Cattle Stomp Shows the Benefit of Healing Hooves," *BEEF Magazine*, May 7, 2014.

Chapter 6: People

1. R. Louv, *Last Child in the Woods: Saving Our Children from Nature Deficit Disorder* (Algonquin Books, 2006).

2. Louv, *Last Child in the Woods*, 157.

3. Louv, *Last Child in the Woods*, 158.

4. Louv, *Last Child in the Woods*, 132–33.

5. G. Nabhan and S. Trimble, *The Geography of Childhood: Why Children Need Wild Places* (Beacon Press, 1994), 126.

6. Louv, *Last Child in the Woods*, 142–43.

7. Louv, *Last Child in the Woods*, 158.

8. Louv, *Last Child in the Woods*, 288.

9. M. Velasquez-Manoff, "A Cure for the Allergy Epidemic?" *New York Times*, November 10, 2013.

10. Velasquez-Manoff, "A Cure for the Allergy Epidemic?"

11. Velasquez-Manoff, "A Cure for the Allergy Epidemic?"

12. https://well.blogs.nytimes.com/2016/02/11/how-the-dirt-cure-can-make-for -healthier-families/.

13. "Screen Time and Children," a document of the National Institutes of Health (NIH), www.nlm.nih.gov/medlineplus/ency/patientinstructions/000355.htm.

14. See, for instance, R. Nauert, "Childhood Television Watching Correlated to Later Attention Problems," *Psych Central*, September 6, 2007, http://psychcentral.com /news/2007/09/06/childhood-television-watching-correlated-to-later-attention -problems/1238.html.

15. See D. Bell, "Protean Manifestations of Vitamin D Deficiency, Part 1: The Epidemic of Deficiency," *Southern Medical Journal*, 2011, www.medscape.com /viewarticle/742623_2.

16. "Memo to Pediatricians: Screen All Kids for Vitamin D Deficiency," website of the Johns Hopkins Children's Hospital, February 22, 2012, www.hopkinschildrens .org/Screen-All-Kids-for-Vitamin-D-Deficiency.aspx.

17. "Vitamin D Deficiency: A Real Problem," website of UC Berkeley Health Services, http://uhs.berkeley.edu/home/healthtopics/pdf/Vitamin%20D%20 Deficiency.pdf.

Chapter 7: Health Claims Against Beef

1. American Public Health Association newsletter, January 22, 2014, http://action. apha.org/site/MessageViewer?dlv_id=49703&em_id=45382.0.

2. R. Lustig, *Fat Chance: Beating the Odds Against Sugar, Processed Food, Obesity, and Disease* (Plume 2012), 4.

3. US Centers for Disease Control and Prevention (CDC) statistics for the most recent year available (2018) show the following as five the leading causes of death: (1) heart disease; (2) cancer; (3) accidents; (4) chronic lower respiratory diseases; and (5) strokes. www.cdc.gov/nchs/fastats/lcod.htm.

4. "Coronary Artery Bypass Graft Surgery Numbers Drop 30% in 7 Years," *Medical News Today*, May 4, 2011, www.medicalnewstoday.com/articles/224100.php.

5. See "Trends in Tobacco Use," www.lung.org/finding-cures/our-research/trend -reports/Tobacco-Trend-Report.pdf; and the website of the CDC, www.cdc.gov /nchs/data/nhis/earlyrelease/earlyrelease201306_08.pdf.

6. B. Keirnan, "Grass Fed vs. Corn Fed: You Are What Your Food Eats," July 16, 2012, www.globalaginvesting.com/news/blogdetail?contentid=1479.

7. See "Disease Statistics," Fact Book 2012, website of the US Department of Health and Human Services, www.nhlbi.nih.gov/about/factbook/chapter4.htm; and Blodget, H. "What Kills Us: The Leading Causes of Death from 1900–2010," *Business Insider*, June 2012l, www.businessinsider.com/leading-causes-of-death -from-1900-2010-2012-6?op=1.

8. T. Parker-Pope, "Less Active at Work, Americans Have Packed on the Pounds," *New York Times*, May 25, 2011, http://well.blogs.nytimes.com/2011/05/25/less-active-at-work-americans-have-packed-on-pounds/?_php=true&_type=blogs&_r=0.

9. G. Reynolds, "The Couch Potato Goes Global," *New York Times*, July 18, 2012, http://well.blogs.nytimes.com/2012/07/18/the-couch-potato-goes-global.

10. See, for example, R. Brownson and T. Boehmer, "Declining Rates of Physical Activity in the United States: What Are the Contributors?," *Annual Review of Public Health* 26 (April 2005), 8, which shows that "low-activity jobs" increased from 1950 (23 percent) to 1970 (41 percent), but then leveled off. See also Institute of Medicine, "Does the Built Environment Influence Physical Activity?" *Examining the Evidence, Special Report 282* (National Academies Press, 2005), which shows (p. 65) that the Model T was introduced in 1908 and that by 1930 electric dish and clothes washers, dryers, and vacuums had all been introduced; and (p. 73) that the major shift in population distribution had occurred by 1950.

11. M. Pollan, "Unhappy Meals." *New York Times Magazine*, January 28, 2007, http://michaelpollan.com/articles-archive/unhappy-meals.

12. For an example of the public health message advising people to cut out red meat and saturated fats, see Kaiser Permanente newsletter, February 2014.

13. A. Keys, "Coronary Heart Disease in Seven Countries," *Circulation* 41, no. 1 (1970), 211.

14. H. Wells and J. Buzby, "Dietary Assessment of Major Trends in US Food Consumption, 1970–2005," a study by the Economic Research Service (ERS) of the US Department of Agriculture, March 2008; S. Gerrior et al., "Nutrient Content of the US Food Supply, 1909–2000," a study of the Center for Nutrition Policy and Promotion (CNPP), US Department of Agriculture, November 2004.

15. Wells and Buzby, "Dietary Assessment of Major Trends in US Food Consumption, 1970–2005"; Gerrior, "Nutrient Content of the US Food Supply, 1909–2000."

16. Recent food consumption figures from Wells and Buzby, "Dietary Assessment of Major Trends in US Food Consumption, 1970–2005," and Gerrior, "Nutrient Content of the US Food Supply, 1909–2000"; historical consumption figures from Gerrior and the database of the Economic Research Service of the US Department of Agriculture, http://ers.usda.gov/data-products/food-consumption-and-nutrient-intakes.aspx#.U6m0mqjc0Xw.

17. M. Enig, *Know Your Fats: The Complete Primer for Understanding the Nutrition of Fats, Oils, and Cholesterol* (Bethesda Press, 2000), 93.

18. ERS/CNPP/USDA data.

19. ERS/CNPP/USDA data.

20. "Per Capita Consumption," a report of the US National Oceanic and Atmospheric Administration (NOAA) Office of Science and Technology, National Marine Fisheries Service, www.st.nmfs.noaa.gov/Assets/commercial/fus/fus99/per_capita99.pdf.

21. D. Maron, "Some Danish Advice on the Trans Fat Ban," *Scientific American*, November 14, 2013, https://www.scientificamerican.com/article/some-danish-advice-on-the/.

22. "FDA Targets Trans Fat in Processed Foods," news release of the US Food and Drug Administration, November 7, 2013, www.fda.gov/ForConsumers /ConsumerUpdates/ucm372915.htm.

23. M. Warner, "A Lifelong Fight Against Trans Fat," *New York Times*, December 16, 2013, www.nytimes.com/2013/12/17/health/a-lifelong-fight-against-trans-fat. html?smid=fb-share&_r=0.

24. Ibid.

25. A. Sachdeva et al., "Lipid Levels in Patients in Hospitalized with Coronary Artery Disease," *American Heart Journal*, January 2009, www.ahjonline.com /article/S0002-8703%2808%2900717-5/abstract.

26. "Most Heart Attack Patients' Cholesterol Level Did Not Indicate Cardiac Risk," https://www.sciencedaily.com/releases/2009/01/090112130653.htm.

27. J. Yudkin, *Pure, White, and Deadly* (Penguin Books, 2013), chapter 2, 81.

28. For more about what Keys chose to exclude, see Keith, *The Vegetarian Myth*, chapter 4.

29. C. Kresser, "Red Meat: It Does a Body Good!," http://chriskresser.com/red-meat -it-does-a-body-good.

30. D. Freedman, "Lies, Damned Lies, and Medical Science," *The Atlantic*, November 2010.

31. Freedman, "Lies, Damned Lies, and Medical Science."

32. https://med.stanford.edu/news/all-news/2018/07/john-ioannidis-calls-for-more -rigorous-nutrition-research.html, July 16, 2018 (emphasis added).

33. U. Ravnskov, "The Questionable Role of Saturated and Polyunsaturated Fatty Acids in Cardiovascular Disease," *Journal of Clinical Epidemiology* 51, no. 6 (June 1998), 443–60, www.ncbi.nlm.nih.gov/pubmed/9635993.

34. P. W. Siri-Tarino et al. "Meta-Analysis of Prospective Cohort Studies Evaluating the Association of Saturated Fat with Cardiovascular Disease," *American Journal of Clinical Nutrition*, March 2010, www.ncbi.nlm.nih.gov/pubmed/20071648.

35. R. Micha, S. Wallace, and D. Mozaffarian, "Red and Processed Meat Consumption and Risk of Incident Coronary Heart Disease, Stroke, and Diabetes Mellitus: A Systematic Review and Meta-Analysis," *Circulation*, May 17, 2010, www.hsph .harvard.edu/news/press-releases/processed-meats-unprocessed-heart-disease-diabetes.

36. A. O'Connor, "Study Questions Fat and Heart Disease Link," *New York Times*, March 17, 2014, http://well.blogs.nytimes.com/2014/03/17/ study-questions-fat-and-heart-disease-link/?_php=true&_type=blogs&nl=todays- headlines&emc=edit_th_20140318&_r=0.

37. N. Teicholz, *The Big Fat Surprise: Why Butter, Meat and Cheese Belong in a Healthy Diet* (Simon and Schuster, 2014), 330–31.

38. H. McGee, *On Food and Cooking: The Science and Lore of the Kitchen*, rev. ed. (Scribner, 2004), 124. Note, too, that marinating meat can substantially reduce the creation of the worrisome compounds. According to Chris Kresser, "If you do want to grill or fry your meats, you can significantly reduce the formation of all of these compounds by using an acidic marinade, which has the added bonus of tasting great! Marinating beef for one hour reduced AGE formation by over

half, and marinades can cut HA formation in meat by up to 90%." See: http://
chriskresser.com/does-red-meat-cause-inflammation.

39. R. Champeau, "Most Heart Attack Patients' Cholesterol Levels Did Not Indicate
    Cardiac Ris," *UCLA Newsroom*, January 12, 2009, http://newsroom.ucla.edu
    /portal/ucla/majority-of-hospitalized-heart-75668.aspx.

40. Mensink et al., "Effects of Dietary Fatty Acids and Carbohydrates on the Ratio
    of Serum Total to HDL Cholesterol and on Serum Lipids and Apolipoproteins:
    A Meta-Analysis of 60 Controlled Trials," *American Journal of Clinical Nutrition*,
    May 2003, www.ncbi.nlm.nih.gov/pubmed/12716665?dopt=Citation.

41. See O'Connor, "Study Questions Fat and Heart Disease Link," and C. Kresser,
    "The Diet–Heart Myth: Cholesterol and Saturated Fat Are Not the Enemy,"
    http://chriskresser.com/the-diet-heart-myth-cholesterol-and-saturated-fat-are
    -not-the-enemy.

42. O'Connor, "Study Questions Fat and Heart Disease Link"; Kresser, "The Diet–
    Heart Myth."

43. D. Mozaffarian, "The Optimal Diet to Prevent CVD [Cardiovascular Disease]:
    What Is the Role of Saturated Fat?," a plenary talk at the Health Effects of
    Dietary Fatty Acids Symposium, Wayne State University, Detroit, Michigan,
    2010, www.meandmydiabetes.com/wp-content/uploads/2010/11/Feinman
    -Webinar-Mozaffarian-Part-1.mp3.

44. Yudkin's clinical experiments and their results are described in detail in chapter 14
    of his book *Pure, White, and Deadly*.

45. Yudkin, *Pure, White, and Deadly*, chapter 14.

46. Yudkin, *Pure, White, and Deadly*, 172.

47. Yudkin, *Pure, White, and Deadly*, 172.

48. Yudkin, *Pure, White, and Deadly*, 172.

49. J. Smith, "John Yudkin: The Man Who Tried to Warn Us About Sugar," *Daily
    Telegraph*, February 17, 2014, https://www.telegraph.co.uk/lifestyle/wellbeing
    /diet/10634081/John-Yudkin-the-man-who-tried-to-warn-us-about
    -sugar.html.

50. Smith, "The Man Who Tried to Warn Us About Sugar."

51. Yudkin, *Pure, White, and Deadly*, 188.

52. Yudkin, *Pure, White, and Deadly*, 113, 115.

53. Yudkin, *Pure, White, and Deadly*, 188.

54. Smith, "The Man Who Tried to Warn Us About Sugar."

55. *Nutrition Source*, website of the Harvard School of Public Health, citing T. T.
    Fung et al., "Sweetened Beverage Consumption and Risk of Coronary Heart
    Disease in Women," *American Journal of Clinical Nutrition* 89 (2009), 1037, www
    .hsph.harvard.edu/nutritionsource/healthy-drinks/soft-drinks-and-disease.

56. *Nutrition Source*, website of the Harvard School of Public Health, citing L. de
    Koning, "Sweetened Beverage Consumption, Incident Coronary Heart Disease,
    and Biomarkers of Risk in Men," *Circulation* 125 (2012), 1735.

57. V. S. Malik, "Sugar-Sweetened Beverages and Risk of Metabolic Syndrome and
    Type 2 Diabetes: A Meta-Analysis," *Diabetes Care* 33 (2010), 2477.

58. "Soft Drinks and Disease," *Nutrition Source*, website of the Harvard School of Public Health, www.hsph.harvard.edu/nutritionsource/healthy-drinks/soft -drinks-and-disease.

59. L. Schmidt, "New Unsweetened Truths About Sugar," *Journal of the American Medical Association Internal Medicine*, April 2014, http://archinte.jamanetwork .com/article.aspx?articleid=1819571.

60. See "Added Sugar Linked to Cardiovascular Disease," www.foodconsumer.org /newsite/2/19/sugar_heart_disease_0302140949.html.

61. M. Busko, "A Soda a Day Ups CVD Risk by 30%: NHANES Study," *Medscape*, February 4, 2014, www.medscape.com/viewarticle/820172.

62. Q. Yang et al., "Added Sugar Intake and Cardiovascular Diseases Mortality Among US Adults," *Journal of the American Medical Association Internal Medicine*, April 2014, http://archinte.jamanetwork.com/article.aspx?articleid=1819573.

63. M. Hyman, "Fat Does Not Make You Fat," *Huffington Post*, November 26, 2013, www.huffingtonpost.com/dr-mark-hyman/fat-health_b_4343798.html.

64. J. Beilby, "Definition of Metabolic Syndrome: Report of the National Heart, Lung, and Blood Institute / American Heart Association Conference on Scientific Issues Related to Definition," *Circulation* 109 (2004), 433, www.ncbi.nlm.nih.gov/ pmc/articles/PMC1880831.

65. Beilby, "Definition of Metabolic Syndrome."

66. A. Malhotra, "Sugar Is Now Enemy Number One," *Guardian*, January 11, 2014, www .theguardian.com/commentisfree/2014/jan/11/sugar-is-enemy-number-one-now.

67. C. Paddock, "Link Between Saturated Fat and Heart Disease Questioned," Medical News Today, May 28, 2015, https://www.medicalnewstoday.com/articles/274166.

68. Beilby, "Definition of Metabolic Syndrome."

69. K. Musunuru, "Atherogenic Dyslipidemia: Cardiovascular Risk and Dietary Intervention," *Lipids* 45, no. 10 (October 2010), www.ncbi.nlm.nih.gov/pmc/articles /PMC2950930.

70. Musunuru, "Atherogenic Dyslipidemia."

71. Musunuru, "Atherogenic Dyslipidemia."

72. H. Briggs, "WHO: Daily Sugar Intake 'Should Be Halved,'" *BBC News*, March 5, 2014, www.bbc.com/news/health-26449497.

73. L. Te Morenga, "Dietary Sugars and Body Weight: Systematic Review and Meta-Analyses of Randomised Controlled Trials and Cohort Studies," *British Medical Journal*, 2013, www.bmj.com/content/346/bmj.e7492.

74. Yang, "Added Sugar Intake and Cardiovascular Diseases Mortality."

75. Lustig, *Fat Chance*, 7 (emphasis added).

76. R. Lustig, "The Bitter Truth," a talk at UCSF, University of California television, uploaded July 30, 2009, www.youtube.com/watch?v=dBnniua6-oM.

77. Smith, "The Man Who Tried to Warn Us About Sugar."

78. "Is Sugar Toxic? *60 Minutes* Investigates." *60 Minutes* television segment, January 2013, https://www.youtube.com/watch?v=6n29ZIJ-jQA.

79. R. Johnson et al., "Potential Role of Sugar (Fructose) in the Epidemic of Hypertension, Obesity, and the Metabolic Syndrome, Diabetes, Kidney Disease,

DEFENDING BEEF

and Cardiovascular Disease," *American Journal of Clinical Nutrition*, October 2007, http://ajcn.nutrition.org/content/86/4/899.long#sec-3.

80. Johnson, "Potential Role of Sugar (Fructose) in the Epidemic" (emphasis added).
81. Yudkin, *Pure, White, and Deadly*, 42.
82. On February 7, 2021, Walmart listed its price for a 5-pound bag of sugar as $3.58.
83. "World War II Rationing," *U.S. History* (website), www.u-s-history.com/pages/h1674.html.
84. R. Sicotte, "The Origins and Development of the US Sugar Program, 1934–1959," paper prepared for the 14th International Economic History Conference, August 2006, www.uvm.edu/~rsicotte/US%20Sugar%20Program.pdf.
85. All statistics from the United Nations Food and Agriculture Organization, from FAOSTAT at faostat.fao.org, and from the FAO report "Food Outlook: Global Market Analysis," May 2012, 41, www.fao.org/docrep/015/al989e/al989e00.pdf.
86. Johnson, "Potential Role of Sugar (Fructose) in the Epidemic."
87. S. Nielsen, "Changes in Beverage Intake Between 1977 to 2001," *American Journal of Preventive Medicine* 27 (October 2004), 205, www.ajpmonline.org/article/S0749-3797%2804%2900122-9/abstract.
88. "USC Research Finds Sodas Sweetened with More High Fructose Corn Syrup than Previously Assumed," press release of the Keck School of Medicine at USC, October 27, 2010, http://keck.usc.edu/en/About/Administrative_Offices/Office_of_Public_Relations_and_Marketing/News/Detail/archive__offices__public_relations_and_marketing__high_fructose_corn_syrup_in_sodas.
89. "USC Research Finds Sodas Sweetened with More High Fructose Corn Syrup."
90. De Koning, "Sweetened Beverage Consumption."
91. Also see a 2007 study published in the *New England Journal of Medicine* that connected HFCS to a variety of health problems and illnesses, including obesity, cardiovascular disease, and renal disease: www.jwatch.org/pa200712120000005/2007/12/12/high-fructose-corn-syrup-new-trans-fat.
92. "USC Research Finds Sodas Sweetened with More High Fructose Corn Syrup."
93. S. Lakhan, "The Emerging Role of Dietary Fructose in Obesity and Cognitive Decline," *Nutrition Journal* 12 (2013), 114, www.nutritionj.com/content/12/1/114.
94. N. Melville, "Culinary Culprits: Foods That May Harm the Brain," *Medscape*, January 30, 2014, www.medscape.com/viewarticle/819974#3.
95. R. Lustig, "The Sugar-Addiction Taboo," *The Atlantic*, January 2014, www.theatlantic.com/health/archive/2014/01/the-sugar-addiction-taboo/282699.
96. N. Avena et al., "Evidence for Sugar Addiction: Behavioral and Neurochemical Effects of Intermittent, Excessive Sugar Intake," *Neuroscience Behavior Review* 32, no. 1 (2008), www.ncbi.nlm.nih.gov/pubmed/17617461.
97. Stice et al., "Relative Ability of Fat and Sugar Tastes to Activate Reward, Gustatory, and Somatosensory Regions," *Journal of Clinical Nutrition* 98 (December 2013), 1377, www.ncbi.nlm.nih.gov/pubmed/24132980.
98. Lustig, *Fat Chance*, 62.
99. L. Sun, "A Mother's Food Choice Can Shape Baby's Palate, Research Shows," *Washington Post*, November 17, 2011, www.washingtonpost.com/national

— 268 —

/health-science/a-mothers-food-choice-can-shape-babys-palate-research-shows/2011/11/17/gIQAqcYooN_story.html.

100. S. Fomon, "Infant Feeding in the 20th Century," *Journal of Nutrition* 131 (February 1, 2001), http://jn.nutrition.org/content/131/2/409S.full#FN1.

101. Lustig, *Fat Chance*, 12.

102. Institute of Medicine, *Nutrition During Lactation* (National Academies Press, 1991), 118, www.nap.edu/openbook.php?record_id=1577&page=118; see also regarding the nutritional components of breast milk, www.parentingscience.com /calories-in-breast-milk.html.

103. J. Moskin, "For an All-Organic Formula, Baby, That's Sweet," *New York Times*, May 19, 2008, www.nytimes.com/2008/05/19/us/19formula.html?_r=0.

104. N. Terrero, "How Much Sugar Is in Brand-Name Baby Formula?" *NBC Latino*, February 22, 2012, http://nbclatino.com/2012/02/22/18091566837.

105. Moskin, "For an All-Organic Formula, Baby, That's Sweet."

106. B. Gibbs, "Socioeconomic Status, Infant Feeding Practices and Early Childhood Obesity," *Pediatric Obesity* 9 (April 2014), 135, www.ncbi.nlm.nih.gov/ pubmed/23554385.

107. "AAP Reaffirms Breastfeeding Guidelines," website of the American Academy of Pediatrics, February 27, 2012, www.aap.org/en-us/about-the-aap/aap-press-room /Pages/AAP-Reaffirms-Breastfeeding-Guidelines.aspx.

108. M. Bartick et al., "The Burden of Suboptimal Breastfeeding in the United States: A Pediatric Cost Analysis," *Pediatrics* 125 (May 2010), http://pediatrics.aap publications.org/content/125/5/e1048.full.

109. "AAP Reaffirms"; and ERS/CNPP/USDA data.

110. "Carbohydrates and Blood Sugar," *Nutrition Source*, website of the Harvard School of Public Health, www.hsph.harvard.edu/nutritionsource/carbohydrates /carbohydrates-and-blood-sugar.

111. K. Fillion, "On the Evils of Wheat," *Macleans*, September 20, 2011, www.macleans .ca/general/on-the-evils-of-wheat-why-it-is-so-addictive-and-how-shunning-it -will-make-you-skinny.

112. N. Jackson, "Your Addiction to Wheat Products Is Making You Fat and Unhealthy," *The Atlantic*, September 22, 2011, www.theatlantic.com/health/archive/2011/09/your -addiction-to-wheat-products-is-making-you-fat-and-unhealthy/245526.

113. Mercola, "Avoid This Food to Help You Slow Aging," February 22, 2012, http://articles .mercola.com/sites/articles/archive/2012/02/22/how-sugar-accelerates-aging.aspx.

114. J. Hamblin, "This Is Your Brain on Gluten," *The Atlantic*, December 20, 2013, www. theatlantic.com/health/archive/2013/12/this-is-your-brain-on-gluten/282550.

115. ERS/CNPP/USDA data.

116. G. Taubes, "What If It's All Been a Big Fat Lie?," *New York Times Magazine*, July 7, 2002, www.nytimes.com/2002/07/07/magazine/what-if-it-s-all-been-a-big-fat -lie.html?scp=1&sq=gary%20taubes%20and%20fat&st=cse.

117. "Carbohydrates," *Nutrition Source*, website of the Harvard School of Public Health, January 2014, www.hsph.harvard.edu/nutritionsource/carbohydrate -question/?utm_source=SilverpopMailing&utm_medium=.

118. C. Gardner et al., "Comparison of the Atkins, Zone, Ornish, and LEARN Diets for Change in Weight and Related Risk Factors Among Overweight Premenopausal Women: The A to Z Weight Loss Study: A Randomized Trial," *Journal of the American Medical Association* 297 (March 7, 2007), http://jama.jamanetwork.com/article.aspx?articleid=205916.

119. "Stanford Diet Study Tips Scale in Favor of Atkins Plan, Stanford Research on Weight Loss," press release of Stanford University, March 1, 2007, http://nutrition.stanford.edu/documents/AZ_press.pdf.

120. Y. Song et al., "A Prospective Study of Red Meat Consumption and Type 2 Diabetes in Middle-Aged and Elderly Women, the Women's Health Study," *Diabetes Care* 27 (September 2004), http://care.diabetesjournals.org/content/27/9/2108.full.

121. S. Ley, "Associations Between Red Meat Intake and Biomarkers of Inflammation and Glucose Metabolism in Women," *American Journal of Clinical Nutrition*, February 2014, http://ajcn.nutrition.org/content/early/2013/11/27/ajcn.113.075663.abstract.

122. C. Kresser, "Does Eating Red Meat Increase the Risk of Diabetes?," chriskresser.com/does-red-meat-cause-inflammation-and-impaired-glucose-metabolism.

123. M. Sisson, "Does Eating Red Meat Increase Type 2 Diabetes Risk?," www.marksdailyapple.com/does-eating-red-meat-increase-type-2-diabetes-risk/#ixzz2xZj5DoBh.

124. "Cleveland Clinic Researchers Discover Link Between Heart Disease and Compound Found in Red Meat, Energy Drinks," news release of Cleveland Clinic, April 7, 2013, http://my.clevelandclinic.org/media_relations/library/2013/2013-04-07-cleveland-clinic-researchers-discover-link-between-heart-disease-and-compound-found-in-red-meat-energy-drinks.aspx.

125. D. Pendrick, "New Study Links L-Carnitine in Red Meat to Heart Disease," *Harvard Health Publications*, April 17, 2013, www.health.harvard.edu/blog/new-study-links-l-carnitine-in-red-meat-to-heart-disease-201304176083.

126. Regarding a vegetarian diet not being optimal, see N. Burkert et al., "Nutrition and Health—The Association Between Eating Behavior and Various Health Parameters: A Matched Sample Study," *PLOS ONE* 9, no. 2 (February 2014), www.plosone.org/article/fetchObject.action?uri=info:doi/10.1371/journal.pone.0088278&representation=PDF.

127. C. Kresser, "Red Meat and TMAO: Cause for Concern, or Another Red Herring?," http://chriskresser.com/red-meat-and-tmao-its-the-gut-not-the-meat.

128. C. Kresser, "Choline and TMAO: Eggs Still Don't Cause Heart Disease," http://chriskresser.com/choline-and-tmao-eggs-still-dont-cause-heart-disease, citing C. de Filippo, "Impact of Diet in Shaping Gut Microbiota Revealed by a Comparative Study in Children from Europe and Rural Africa," *Proceedings of the National Academy of Sciences*, August 17, 2010, www.ncbi.nlm.nih.gov/pmc/articles/PMC2930426.

129. M. Janeiro et al., "Implication of Trimethylamine N-Oxide (TMAO) in Disease: Potential Biomarker or New Therapeutic Target," *Nutrients*, October 2018, https://pubmed.ncbi.nlm.nih.gov/?term=Janeiro+MH&cauthor_id=30275434.

130. L. Stowkowski, "Red Meat and Cancer: What's the Beef?" *Medscape*, June 20, 2013, www.medscape.com/viewarticle/806573.

131. S. Rohrmann et al., "Meat Consumption and Mortality—Results from the European Prospective Investigation into Cancer and Nutrition," *BMC Medicine* 11 (2013), 63, www.biomedcentral.com/1741-7015/11/63.

132. D. Alexander et al., "Meta-Analysis of Prospective Studies of Red Meat Consumption and Colorectal Cancer," *European Journal of Cancer Prevention* 20, no. 4 (July 2011), 293, www.ncbi.nlm.nih.gov/pubmed/21540747; see also A. Truswell, "Meat Consumption and Cancer of the Large Bowel," *European Journal of Clinical Nutrition*, 2002, www.ncbi.nlm.nih.gov/pubmed/11965518.

133. S. Missmer et al., "Meat and Dairy Food Consumption and Breast Cancer: A Pooled Analysis of Cohort Studies," *International Journal of Epidemiology* 31, no. 1 (February 2002), 78, www.ncbi.nlm.nih.gov/pubmed/11914299.

134. "Dairy Products and Breast Cancer Risk: A Review of the Literature" *International Journal of Fertility and Women's Medicine* 50, no. 6 (November 2005), 244–49, https://www.researchgate.net/publication/7252770.

135. "Scientific Advances of Milk Enrichment with Conjugated Linoleic Acid to Produce Anti-Cancer Milk," https://sites.kowsarpub.com/ijcm/articles/5868.html.

136. K. Clancy, *Greener Pastures: How Grass-Fed Beef and Milk Contribute to Healthy Eating*, a report of the Union of Concerned Scientists, 2006, 1, www.ucsusa.org /assets/documents/food_and_agriculture/greener-pastures.pdf.

137. F. Kolahdooz, "Meat, Fish, and Ovarian Cancer Risk: Results from 2 Australian Case-Control Studies, a Systematic Review, and Meta-Analysis," *American Journal of Clinical Nutrition* 91, no. 6 (June 2010), 1752, www.ncbi.nlm.nih.gov /pubmed/20392889.

138. S. Rohrmann, "Meat and Fish Consumption and Risk of Pancreatic Cancer: Results from the European Prospective Investigation into Cancer and Nutrition," *International Journal of Cancer* 132, no. 3 (2013), 617, www.medscape.com /medline/abstract/22610753.

139. "Unprocessed Red Meat and Processed Meat Consumption: Dietary Guideline Recommendations from the Nutritional Recommendations (NutriRECS) Consortium," *Annals of Internal Medicine*, 2019, https://www.acpjournals.org /doi/10.7326/M19-1621; and https://www.nytimes.com/2019/09/30/health /red-meat-heart-cancer.html.

140. Mozaffarian, "The Optimal Diet to Prevent CVD."

141. "What Are Telomeres?," https://www.news-medical.net/life-sciences/What-are -Telomeres.aspx.

142. https://www.sciencealert.com/study-links-ultra-processed-junk-food-to-age-marker -in-chromosomes?fbclid=IwAR33TrgGPLildk8dRk6mQuFj0bBhOwf6d- pYA0HdRcdXbLd1THuXngx4RpMI; and *American Journal of Clinical Nutrition* 111, no. 6 (June 2020), 1259–66, https://doi.org/10.1093/ajcn/nqaa075.

143. E. Atkins, "The Promise and Problem of Fake Meat," *The New Republic*, June 7, 2019.

144. M. Simon, "Lab-Grown Meat Is Coming, Whether You Like It or Not," *Wired Magazine*, 2018.

145. M. Gurven and H. Kaplan, "Hunter Gatherer Health," a faculty paper, 2007, www.anth.ucsb.edu/faculty/gurven/papers/GurvenKaplan2007pdr.pdf (internal citations omitted).

146. "An Empirical Study of Chronic Diseases in the United States," *International Journal of Environmental Research and Public Health*, March 2018, https://www.ncbi.nlm.nih.gov/pmc/articles/PMC5876976/.

147. "An Empirical Study of Chronic Diseases in the United States" (internal citations omitted).

148. S. Lin, *The Dental Diet*, 2nd ed. (Hay House, 2018), 38.

149. McGee, *On Food and Cooking*, 124.

150. McGee, *On Food and Cooking*, 124.

151. S. Fallon and M. Enig, "Guts and Grease: The Diet of Native Americans," website of the Weston A. Price Foundation, January 1, 2000, www.westonaprice.org/traditional-diets/guts-and-grease.

152. Fallon and Enig, "Guts and Grease." See also V. Smil, *Should We Eat Meat?* (Wiley, 2013), especially pages 46–49, in which the author, world-renowned biologist Vaclav Smil, explains in detail how hunting preferences were made by early humans, and notes that "the preference of all hunters for killing fatty animals is indisputable. This is why megaherbivores were always preferred, as greater (and often fatal) risks associated with their hunting were trumped by superior energy returns."

153. Fallon and Enig, "Guts and Grease."

154. P. Gadsby and L. Steel, "The Inuit Paradox: How Can People Who Gorge on Fat and Rarely See a Vegetable Be Healthier than We Are?" *Discover Magazine*, October 1, 2004, http://discovermagazine.com/2004/oct/inuit-paradox#.Uuwg KPYhYVk.

155. Fallon and Enig, "Guts and Grease."

156. D. Martin and A. Goodman, "Health Conditions Before Columbus: A Paleopathology of Native North Americans," *Western Journal of Medicine* 176 (January 2002), 65, www.ncbi.nlm.nih.gov/pmc/articles/PMC1071659.

157. I. Petersen, "The Maasai Keep Healthy Despite a High Fat Diet," *Science Nordic*, September 11, 2012, http://sciencenordic.com/maasai-keep-healthy-despite -high-fat-diet.

158. C. Masterjohn, "The Masai Part II: A Glimpse of the Masai Diet at the Turn of the 20th Century—A Land of Milk and Honey, Bananas from Afar," website of the Weston A. Price Foundation, September 13, 2011, www.westonaprice.org /blogs/cmasterjohn/2011/09/13/the-masai-part-ii-a-glimpse-of-the-masai-diet -at-the-turn-of-the-20th-century-a-land-of-milk-and-honey-bananas-from-afar.

159. See also G. Mann et al., "Cardiovascular Disease in the Masai," *Journal of Atherosclerosis Research* 4 (July 8, 1964), 289, www.sciencedirect.com/science /article/pii/S0368131964800417.

160. S. Fallon, *Nourishing Traditions* (Newtrends Publishing, 2003), preface.

161. J. Diamond, *Guns, Germs, and Steel: The Fates of Human Societies* (W. W. Norton, 2009), 112.

162. "Food Prices and Spending," Economic Research Service, https://www.ers
.usda.gov/data-products/ag-and-food-statistics-charting-the-essentials/food
-prices-and-spending/.

Chapter 8: Beef Is Good Food

1. McGee, *On Food and Cooking*, 123.
2. C. Kresser, "The Truth About Red Meat," http://chriskresser.com/the-truth
-about-red-meat.
3. B. Pobiner, "Evidence for Meat-Eating by Early Humans," *Nature Education
Knowledge*, 2013, www.nature.com/scitable/knowledge/library/evidence-for-meat
-eating-by-early-humans-103874273.
4. "Iron: Dietary Supplement Fact Sheet," a National Institutes of Health fact sheet,
reviewed April 8, 2014, http://ods.od.nih.gov/factsheets/Iron-HealthProfessional.
5. "Supplements: Nutrition in a Pill?," website of the Mayo Clinic, January 19, 2013,
www.mayoclinic.org/healthy-living/nutrition-and-healthy-eating/in-depth
/supplements/art-20044894.
6. M. Nestle, *What to Eat* (North Point Press, 2007), 468.
7. Fallon, *Nourishing Traditions*, 27.
8. Fallon, *Nourishing Traditions*, 27.
9. J. Blythman and R. Sykes, "Why Beef Is Good for You," *Guardian*,
December 13, 2013, www.theguardian.com/lifeandstyle/2013/dec/16/
why-beef-is-good-for-you-grass-fed-grain-fed.
10. "Beef, Nutrition Facts," *SELF Nutrition Data: Know What You Eat*, http://
nutritiondata.self.com/facts/beef-products/3477/2.
11. "Nutrition: Micronutrient Deficiencies, Iron Deficiency Anaemia," website of the
World Health Organization, 2014, www.who.int/nutrition/topics/ida/en.
12. A. Meister, "The Role of Beef in the American Diet," a report prepared for the
American Council on Science and Health, January 2003, 10, www.meat-ims.org/wp
-content/uploads/2013/07/The_Role%20_of_Beef_in_the_American_Diet.pdf.
13. "Iron: Dietary Supplement Fact Sheet."
14. "Iron: Dietary Supplement Fact Sheet."
15. "Anemia Fact Sheet," Office on Women's Health, US Department of Health and
Human Services, updated July 16, 2012, http://womenshealth.gov/publications
/our-publications/fact-sheet/anemia.html.
16. "Increasing Iron in Your Diet During Pregnancy," website of the Cleveland Clinic,
updated December 21, 2009, http://my.clevelandclinic.org/healthy_living
/pregnancy/hic_increasing_iron_in_your_diet_during_pregnancy.aspx.
17. See, e.g. https://my.clevelandclinic.org/health/drugs/12871-iron-in-your-diet.
18. "Iron: Dietary Supplement Fact Sheet."
19. "Increasing Iron in Your Diet During Pregnancy."
20. M. Turgeon, *Clinical Hematology: Theory and Procedures* (LWW Publishing,
2011), 133.
21. McGee, *On Food and Cooking*, 134.
22. "Iron: Dietary Supplement Fact Sheet."

23. Turgeon, *Clinical Hematology*, 133.
24. "Zinc: Fact Sheet for Consumers," a National Institutes of Health fact sheet, reviewed September 20, 2011, http://ods.od.nih.gov/factsheets/Zinc-QuickFacts.
25. World Health Organization, *The World Health Report*, https://www.who.int /whr/2002/chapter4/en/index3.html.
26. Meister, "The Role of Beef in the American Diet."
27. "Zinc," chapter 16 in *Human Vitamin and Mineral Requirements*, a joint report of the United Nations Food and Agriculture Organization and the World Health Organization, 2001, www.fao.org/docrep/004/y2809e/y2809e0m.htm.
28. "Animal Protein Intake Is Associated with Higher-Level Functional Capacity in Elderly Adults: The Ohasama Study," *Journal of the American Geriatrics Society* 62, no. 3 (March 2014), 426, http://onlinelibrary.wiley.com/doi/10.1111/jgs.12690 /abstract;jsessionid=B96850F343D96439D7E05E97E4DC7CEC.f03t02?sys- temMessage=Wiley+Online+Library+will+be+disrupted+Saturday%2C+15+- March+from+10%3A00-12%3A00+GMT+%2806%3A00-08%3A00+EDT%29+- for+essential+maintenance. Also note: Even a March 2014 study widely reported in the media as showing that middle-aged men should eat less meat actually concluded that older people who ate more meat were healthier. www.cell.com/ cell-metabolism/abstract/S1550-4131%2814%2900062-X.
29. G. Reynolds, "Lift Weights, Eat More Protein, Especially if You're Over 40," New York Times, February 7, 2018, https://www.nytimes.com/2018/02/07/well /move/lift-weights-eat-more-protein-especially-if-youre-over-40.html?auth=login -email&login=email.
30. R. Heaney and D. Layman, "Amount and Type of Protein Influences Bone Health," *American Journal of Clinical Nutrition* 87, no. 5 (May 2008), http://ajcn .nutrition.org/content/87/5/1567S.full.
31. "Animal Protein Is Good for Bones," website of Hebrew SeniorLife, Harvard Medical School affiliate, www.hebrewseniorlife.org/research-animal-protein -is-good-for-bones.
32. M. Nelson, *Strong Women, Strong Bones* (The Berkeley Publishing Group, 2000), 4.
33. Nelson, *Strong Women, Strong Bones*, 288.
34. K. Rheaume-Bleue, *Vitamin K₂ and the Calcium Paradox* (Wiley, 2012), 228.
35. F. Gomez-Pinilla, "Brain Foods: The Effects of Nutrients on Brain Function," *Nature Reviews Neuroscience* 9 (July 2008), 568, www.nature.com/nrn/journal/v9 /n7/fig_tab/nrn2421_T1.html.
36. Fallon, *Nourishing Traditions*, 27; Enig, *Know Your Fats*, 57.
37. Fetal and Neonatal Cholesterol Metabolism, https://www.ncbi.nlm.nih.gov /books/NBK395580/.
38. Clancy, *Greener Pastures*.
39. A. O'Connor, "Low Vitamin D Levels Linked to Disease in Two Big Studies," *New York Times*, April 1, 2014, http://well.blogs.nytimes.com/2014/04/01/low -vitamin-d-levels-linked-to-disease-in-two-big-studies/?_php=true&_type= blogs&_r=0.
40. O'Connor, "Low Vitamin D Levels Linked to Disease in Two Big Studies."

41. Kresser, "Red Meat: It Does a Body Good."
42. Kresser, "Red Meat: It Does a Body Good."
43. Kresser, "Red Meat: It Does a Body Good."
44. P. Weintraub, "All About B Vitamins," reprinted from *Experience Life Magazine*, website of Dr. Frank Lipman, www.drfranklipman.com/all-about-b-vitamins.
45. Weintraub, "All About B Vitamins."
46. Weintraub, "All About B Vitamins."
47. Weintraub, "All About B Vitamins"; also see, for instance, Burt Berkson, PhD, MD, director of the Integrative Medicine Center of New Mexico in Las Cruces and coauthor of *User's Guide to the B-Complex Vitamins* (Basic Health Publications, 2005), who writes that most of us are woefully deficient. "Most Americans could probably benefit from taking up to three times—or more—of the RDAs for many of the B vitamins."
48. Meister, "The Role of Beef in the American Diet," 11.
49. "Dietary Supplement Fact Sheet, Vitamin B₆," a National Institutes of Health (NIH) fact sheet, reviewed September 15, 2011, http://ods.od.nih.gov/factsheets /VitaminB6-HealthProfessional.
50. A. Gilsing et al., "Serum Concentrations of Vitamin B₁₂ and Folate in British Male Omnivores, Vegetarians and Vegans: Results from a Cross-Sectional Analysis of the EPIC-Oxford Cohort Study," *European Journal of Clinical Nutrition* 64, no. 9 (September 2010), 933, www.ncbi.nlm.nih.gov/pubmed/20648045.
51. S. Kirchheimer, "Vegetarian Diet and B₁₂ Deficiency," https://www.webmd.com /food-recipes/news/20030618/vegetarian-diet-b12-deficiency#1.
52. Weintraub, "All About B Vitamins."
53. "Vitamin B₁₂, Fact Sheet for Consumers," a National Institutes of Health (NIH) fact sheet, reviewed June 24, 2011, http://ods.od.nih.gov/factsheets /VitaminB12-QuickFacts.
54. http://chriskresser.com/natures-most-potent-superfood.
55. See, for instance, Hahn Niman, *Righteous Porkchop*; and N. Hahn Niman, "For Animals, Grass Each Day Keeps Doctors Away," *The Atlantic*, May 2010, www .theatlantic.com/health/archive/2010/05/for-animals-grass-each-day-keeps -doctors-away/56915.
56. McGee, *On Food and Cooking*, 134.
57. McGee, *On Food and Cooking*, 135
58. McGee, *On Food and Cooking*, 121.
59. See, for instance, F. Diez-Gonzalez et al., "Grain Feeding and the Dissemination of Acid-Resistant *Escherichia coli* from Cattle," *Science* 281 (1998), 1666.
60. Clancy, *Greener Pastures*.
61. Clancy, *Greener Pastures*, 50.
62. Clancy, *Greener Pastures*, 25.
63. Clancy, *Greener Pastures*, 48.
64. Clancy, *Greener Pastures*.
65. J. Robinson, "Summary of Important Health Benefits of Grassfed Meats, Eggs and Dairy," www.eatwild.com/healthbenefits.htm#8, citing S. Duckett et al.,

"Effects of Time on Feed on Beef Nutrient Composition," *Journal of Animal Science* 71, no. 8 (1993), 2079.

66. Keirnan, "Grass Fed vs. Corn Fed."

67. S. Duckett et al., "Effects of Winter Stocker Growth Rate and Finishing System on: III. Tissue Proximate, Fatty Acid, Vitamin, and Cholesterol Content," *Journal of Animal Science* 87, no. 9 (June 2009), 2961, www.journalofanimalscience.org /content/87/9/2961.long.

68. C. Daley et al., "A Review of Fatty Acid Profiles and Antioxidant Content in Grass-Fed and Grain-Fed Beef," *Nutrition Journal* 9 (2010), 10, www.nutritionj .com/content/9/1/10.

69. Robinson, "Summary of Important Health Benefits," citing *British Journal of Nutrition* 105 (2011), 80.

70. "Grassfed Cows Produce Healthier Milk," 2018, https://extension.umn.edu /pasture-based-dairy/grass-fed-cows-produce-healthier-milk.

71. "The 'Grass-Fed' Milk Story: Understanding the Impact of Pasture Feeding on the Composition and Quality of Bovine Milk," *Foods*, August 2019, https://www .ncbi.nlm.nih.gov/pmc/articles/PMC6723057/.

72. "The 'Grass-Fed' Milk Story."

73. M. Burros, "There's More to Like About Grassfed Beef," *New York Times*, August 30, 2006, www.nytimes.com/2006/08/30/dining/30well.html.

74. L. Curry, *Pure Beef: An Essential Guide to Artisan Meat with Recipes for Every Cut* (Running Press, 2012), 58.

75. McGee, *On Food and Cooking*, 138.

76. "Tackling Climate Change," FAO.

77. Pimentel and Pimentel, "[A] Feedback Mechanism May Exist," 129.

78. M. Snell, "Medieval Food Preservation," http://historymedren.about.com/od /foodandfamine/a/food_preservation.htm.

79. McGee, *On Food and Cooking*, 145.

80. A. Danforth, *Butchering Beef: The Comprehensive Photographic Guide to Humane Slaughtering and Butchering* (Storey Publishing, 2014), 93.

A Critique: What's the Matter with Beef?

1. "Meat Consumption Dips as Consumers Focus on Healthier Diets: Mintel," *Meat + Poultry*, January 22, 2014, https://www.meatpoultry.com/articles/10005-meat -consumption-dips-as-consumers-focus-on-healthier-diets-mintel.

2. "Cargill's View on Zilmax Being Pulled from the Market," Cargill News Center, www .cargill.com/news/cargill-view-on-zilmax-being-pulled-from-the-market/index.jsp.

3. "Concerns Raised About Using Beta Agonists in Beef Cattle," *Science Daily*, March 12, 2014, www.sciencedaily.com/releases/2014/03/140312181913.htm.

4. T. Grandin, "The Effect of Economics on the Welfare of Cattle, Pigs, Sheep, and Poultry," website of Temple Grandin, April 2013, www.grandin.com/welfare /economic.effects.welfare.html.

5. D. Charles, "Inside the Beef Industry's Battle Over Growth-Promotion Drugs," website of National Public Radio, August 21, 2013, www.npr.org/blogs/thesalt

/2013/08/21/214202886/inside-the-beef-industrys-battle-over-growth
-promotion-drugs.

6. See M. Mellon, *Hogging It: Estimates of Antimicrobial Abuse in Livestock*, a report of the Union of Concerned Scientists, 2001.

7. See, for instance: "It is well established that predators play a vital role in maintaining structure and stability of communities and that removal of predators can have a variety of cascading, indirect effects" (Terborgh et al., 2001; Duffy, 2003). Krausman, "An Assessment of Rangeland Activities."

8. See Hahn Niman, *Righteous Porkchop*, for a more complete discussion of this issue.

9. "Standards," website of Animal Welfare Approved, a program of the Animal Welfare Institute, http://animalwelfareapproved.org/standards/beef-cattle -2014/#130-transport.

Final Analysis: Why Eat Animals?

1. M. Nestle, *Food Politics* (University of California Press, 2002), 13.

2. A. Kimbrell, *The Fatal Harvest Reader: The Tragedy of Industrial Agriculture* (Island Press, 2002), 7.

3 M. McLaughlin, *World Food Security* (Center of Concern, 2002).

4. McLaughlin, *World Food Security*.

5. M. Eisler et al., "Agriculture: Steps to Sustainable Livestock," *Nature* 507 (March 2014), www.nature.com/news/agriculture-steps-to-sustainable-livestock-1.14796.

6. "Tackling Climate Change," FAO, 1.

7. K. Pathak, "Livestock Development: How It Contributes to Smallholder Farms," *Food Tank*, March 6, 2014, http://foodtank.com/news/2014/03/livestock -development-how-it-contributes-to-smallholder-farmers.

8. Fairlie, *Meat*, 109, quoting Dando, *The Geography of Famine* (V. H. Winston and Sons, 1980).

9. McGee, *On Food and Cooking*, 122.

10. Pathak, "Livestock Development."

11. Eisler, "Agriculture: Steps to Sustainable Livestock." Also see, for instance: "[O] xen consume mostly forage, which is unsuitable for human consumption . . . In Guatemala, the use of about 310h of ox power reduces the human labor input almost by half." Pimentel and Pimentel, "[A] Feedback Mechanism May Exist," 103.

12. C. Tudge, *Feeding People Is Easy* (Pari Publishing, 2007), 67.

13. www.rural21.com/english/news/detail/article/livestock-futures-conference -about-powerlessness-and-hope-0000466.

14. Pimentel and Pimentel, "[A] Feedback Mechanism May Exist," 65.

15. Pimentel and Pimentel, "[A] Feedback Mechanism May Exist," 65.

16. Eisler, "Agriculture: Steps to Sustainable Livestock."

17. Eisler, "Agriculture: Steps to Sustainable Livestock."

18. Pimentel and Pimentel, "[A] Feedback Mechanism May Exist," 73.

19. Pimentel and Pimentel, "[A] Feedback Mechanism May Exist," 69.

20. Montgomery, *Dirt*, 35.

21. Pimentel and Pimentel, "[A] Feedback Mechanism May Exist," 62.

22. Pimentel and Pimentel, "[A] Feedback Mechanism May Exist," 62.

23. Troeh, *Soil and Water Conservation*, 337.

24. L. Rinehart, "Ruminant Nutrition for Graziers," National Sustainable Agriculture Information Service, 2008, https://attra.ncat.org/attra-pub/viewhtml.php?id=201.

25. Tudge, *Feeding People Is Easy*, 67.

26. Mooney and Zavaleta, *Ecosystems of California*, (Univ. of California Press, 2016), 2.

27. Mooney and Zavaleta, *Ecosystems of California*.

28. Pimentel and Pimentel, "[A] Feedback Mechanism May Exist," 58, 59.

29. Pimentel and Pimentel, "[A] Feedback Mechanism May Exist," 74.

30. www.beefboard.org/news/files/factsheets/The-Environment-And-Cattle -Production.pdf.

31. Mooney and Zavaleta, *Ecosystems of California*, 2.

32. Mooney and Zavaleta, *Ecosystems of California*, 4.

33. Mooney and Zavaleta, *Ecosystems of California*, 14.

34. Fairlie, *Meat*, 30.

35. Fairlie, *Meat*, 36.

36. N. Hahn Niman, "Can Meat Eaters Also Be Environmentalists?" *The Atlantic*, June 2, 2010, www.theatlantic.com/health/archive/2010/06/can-meat-eaters-also -be-environmentalists/57532.

37. Hahn Niman, "Animals Are Essential."

38. N. Hahn Niman, "Dogs Aren't Dinner: The Flaws in an Argument for Veganism," *The Atlantic*, November 4, 2010, www.theatlantic.com/health/archive/2010/11 /dogs-arent-dinner-the-flaws-in-an-argument-for-veganism/66095.

39. See, for instance, Keith, *The Vegetarian Myth*; and R. Dunn, "Human Ancestors Were Nearly All Vegetarians," *Scientific American*, July 23, 2012, http://blogs .scientificamerican.com/guest-blog/2012/07/23/human-ancestors-were-nearly -all-vegetarians.

40. H. Fearnley-Whittingstall, *The River Cottage Meat Book* (Ten Speed Press, 2007), 12.

41. Fearnley-Whittingstall, *The River Cottage Meat Book*, 16.

42. A. Howard, "The Animal as Our Farming Partner," *Organic Gardening* 2, no. 3 (September 1947), http://journeytoforever.org/farm_library/howard_animal.html.

43. Fairlie, *Meat*, 270.

44. Fairlie, *Meat*, 273.

45. Fairlie, *Meat*, 273.

46. Troeh, *Soil and Water Conservation*, 349.

47. Pimentel and Pimentel, "[A] Feedback Mechanism May Exist," 364.

INDEX

Aberdeen-Angus cattle, 212
absorption of nutrients, 196–197, 198
acorns, 25
Adams, Mark, 20
addiction to sugar, 158, 163
adrenal hormones, 150, 151
age of livestock, 2, 90, 205, 213, 222
aggregation of soil, 35, 37, 61
aging, bone health in, 199, 200
agriculture
    carbon loss in, 10
    chemical-based. *See* chemical-based
      agriculture
    conservation, 54
    cropland in. *See* cropland
    energy use in. *See* energy use
    financial issues in, 115, 116, 226
    fossil fuels in, 74, 76, 85
    greenhouse gases in, 9–10, 22–23, 50
    health changes in advent of, 184, 186
    industrial. *See* industrial agriculture
    lifestyle in, 115–128, 239–240, 245–246
    organic, 31, 32, 114
    pollution in, 72, 77–93
    regenerative. *See* regenerative agriculture
    small-scale farmers in, 225–227
    soil problems in, 63–64, 109–110, 243–244
    subsidies in, 4, 73, 75, 188
Ahrens, Richard, 152–153
alfalfa, 214, 215
algae, 78
Allen, Myles, 14–15
allergies, 126
almonds, 73–74
alpha-linolenic acid (ALA), 206, 208
Alzheimer's disease, 167
American Farm Policy (1948–1973), 188
amino acids, 194, 228
Amish, 126

ammonia, 20, 70, 84
amylopectin A, 166
Anderson, Clinton P., 71
anemia, 186, 195–196, 203
Angus cattle, 212
animal fats. *See* saturated fats
animal feeds, 18–19, 52–53, 96, 221–222
    in confined feeding operations, 79, 82–83
    drugs in, 218–221
    grains in, 205–209, 222, 228, 229
    grass as, 59–76, 204–209
    hay as, 214–215
    seaweed in, 18–19
    soybeans in, 11, 12, 24, 28, 83, 87
animal husbandry, 204–215
    and climate change, 54
    confinement in. *See* confinement operations
    drugs in, 218–221
    enjoyment of, 245–246
    and ethics of meat eating, 235, 236–237,
      241–242
    grass-based, 204–215
    and "it's not the cow, it's the how," 5, 30,
      54, 113
    lifestyle in, 115–128, 181, 239–240
    nature as model for, 241–242
    and quality of food produced, 205, 237
    and taste of meat, 209–215
"Animals Are Essential to Sustainable Food"
    (Niman), 235
antibiotics, 72, 79, 82, 86–87, 204, 205,
    220–221
antioxidants, 208–209
aroma of beef, 189
arthropods, 105
Asparagopsis taxiformis, 19
asthma, 126
atherosclerosis, 143, 149, 150, 153, 157, 166, 173
Atkins diet, 168, 169, 170

bacteria, 19–20, 36, 205, 215

Barber, Dan, 215

beef cattle

    age at slaughter, 2, 90, 205, 213, 222

    breeds of, 212

    carbon footprint of, 49, 84, 181

    care and raising of. *See* animal husbandry

    drugs given to, 218–221, 222

    and farm/ranch life, 115–128, 239–240, 245–246

    in feedlots, 89–91, 204, 218–220, 222, 237

    grass-fed. *See* grass-fed beef

    introduction into US, 69

    long-distance transport of, 222

    methane emissions from, 15, 49

    number in US, 1–2, 3

    pounds of meat per animal, 238

    quality of life, 89, 237

    soy as feed for, 11

    water requirements of, 77, 93–98, 99

beef consumption, 11–12, 24, 131–132

    environmental issues in. *See* environmental issues

    ethical issues in, 224–246

    fat intake in, 131, 137–149

    flavor and taste in, 209–212, 213, 215

    health issues in. *See* health issues

    instincts in, 190

    limitations of research on, 140–142

    in low-carb diets, 168

    moral acceptability of, 224, 234–246

    nutritional value of, 53–54, 96–97, 189–216, 230

    organ meats in, 194, 202, 203, 204

    satisfaction in, 189

    and TMAO production, 172–174

    trends in, 2, 4, 132–133, 134, 218

beef preservation, 215–216

beef tallow, 135

bees, 106

Berry, Wendell, 63

beta-agonist drugs, 218–220

beta-carotene, 207, 208

Beyond Beef (Rifkin), 108

Beyond Meat, 179

The Big Fat Surprise (Teicholz), 144–145

biochar, 114

biodiversity, 30, 41, 43, 75, 81, 100–107

in natural climate solutions, 34

    of plants, 43, 46, 63, 101–102

    of soil organisms, 81, 100

bio-fuels, 73

birds, 104, 105, 106

bison, 17, 25, 66, 68, 69, 112

Biswell, Harold, 25

blood pressure, 130, 131, 137, 156

    sugar affecting, 149, 152, 154, 155, 161

body mass index, 171–172

Bolsonaro, Jair, 10

bone health, 167, 192, 193, 194, 199–200

Bork, Ed, 20

Born to Run (McDougall), 66

Boyd, Kevin, 164

Brahms, Johannes, 65

brain function, 162, 163, 167, 201

Branford, Sue, 11

Brazil, 7, 10–11, 12, 24, 29

breast cancer, 175

breast milk, 164, 165, 197, 201, 203

breeding, 2–3, 88, 212

    grazing of cattle raised for, 90, 91–92, 214, 229, 237

breeds of beef cattle, 212

brittle environment, 43, 70

Brody, Jane, 161

Brown, Ethan, 179

Brown, Gabe, 181

Brown, Jerry, 33

Brown, Pat, 180, 182

Bureau of Land Management, 112

Burkholder, JoAnn, 82

burning practices, historical, 25–26

Burros, Marian, 210

Butchering Beef (Danforth), 216

butter, 1, 2, 4, 131, 133, 192, 200

Caballero, Benjamin, 164

calcium, 199, 200

Call of the Reed Warbler (Massy), 240–241

caloric intake, 132, 158, 160, 161

calves, 212–213, 214, 222

cancer, 130, 167, 174–178

"Can Meat Eaters Also Be Environmentalists?" (Niman), 235

Capra, Fritjof, 58

carbohydrates, 147–170

in beef, 195
in grains, 165–168
in sugar. *See* sugar
carbon, 10
new, in fossil fuel emissions, 17
old, in ruminant emissions, 17
in soil. *See* soil carbon
in trees, 10, 27, 32
carbon cycle, biogenic, 17
carbon dioxide, 9–12, 14–15, 16
in deforestation, 23–24
in grass-based dairy, 83, 84
historical levels of, 65
"The Carnivore's Dilemma" (Niman), 8
cattle, 8–9
beef. *See* beef cattle
in breeding herds, 90, 91–92, 214, 229, 237
in climate change mitigation, 30–51, 54
dairy. *See* dairy cows
digestion process in, 18
domestication of, 68, 127, 236
methane emissions of, 13, 15, 17–20, 28
number in US, 1–2, 3, 13, 91, 111–112
percentage of greenhouse gases attributed
to, 8, 21–23, 24, 28, 29, 52
variety of products from, 29
cellulose, 17, 18, 19, 59
Chan, Margaret, 158
cheap foods, 187–188
cheese, 2, 133, 200
chemical-based agriculture, 20–21, 72, 75–76,
77, 83, 85–86
biodiversity loss in, 75, 101
erosion in, 114
faux meat ingredients from, 181
fossil fuel use in, 74, 85
children
farm experiences of, 116, 117–118, 119,
120, 121–123, 125–128
nature experiences of, 118, 119, 123, 124,
128
overweight and obese, 129
screen time of, 127–128
cholesterol, 136, 146–148, 156, 170, 201
and carbohydrates, 150, 153, 157, 166
and heart disease, 136, 139–140, 146–148,
157
choline, 203

Chowdhury, Rajiv, 144, 147–148
chronic diseases, 2, 130, 137, 138, 139, 177
historical studies of, 183–187
and sugar, 155, 156, 158
and trans fats, 135
Clancy, Kate, 206
clean meat, 178–179, 180
Clean Water Act, 78, 80
Cleveland Clinic, 172–173, 196
climate change, 7–58
carbon dioxide in. *See* carbon dioxide
cattle as factor in, 8, 30
cattle in mitigation of, 30–51, 54
and confinement operations, 21, 22
and default emissions, 28–29
methane in, 9, 12–20, 28, 49, 83, 84
silvopasture in mitigation of, 27–28
transportation as factor in, 8
clothing, and greenhouse gases, 56
Cochrane, Willard, 188
Coleman, Eliot, 51–52
colon cancer, 174–175
colony collapse disorder, 106
Columbus, Christopher, 69
confinement operations, 12, 16, 72, 74,
78–83, 89–91
antibiotics in, 220
dairy cows in, 12, 83, 84, 87–89, 92, 237
ethical issues in, 236–237
pigs and poultry in, 12, 21–22, 28, 78, 79,
80, 82, 220
conjugated linoleic acid, 176, 206–207, 208
Conser, Russ, 5
conservation agriculture, 54
Conservation Reserve Program, 75, 104–105
cooking methods, 136, 146, 174, 210
corn, 73, 83, 85, 90, 91, 188
corn oil, 131, 136, 144
corn syrup, high-fructose, 160–161
Corruccini, Robert, 184
cost and quality of foods, 187–188
cover crops, 26, 31, 70, 110, 234
COVID-19 pandemic, 116
Cows Save the Planet (Schwartz), 39
cropland, 63–64, 83, 109–110, 243–244
biodiversity loss in, 100, 101
compared to grasslands, 111
converted to grassland, 38, 39, 84

cropland (*continued*)
    ecosystem disruption in, 241
    grassland converted to, 38, 61, 69–70, 73
    limitations of grassland as, 230–233
crops
    compared to livestock keeping, 226
    grasslands unsuitable for, 230–233
    grazing on residues, 230
    and livestock integration, 31, 226–227, 242
    pollinators for, 106–107
    rotation of, 21, 31, 36–37, 54, 64, 70
Curry, J. P., 61
Curry, Lynne, 210
cyanocobalimin, 194, 203–204

dairy cows, 2–3, 15, 16, 19, 52, 53
    confinement of, 12, 83, 84, 87–89, 92, 237
    grass-fed, 83–84, 92, 204, 206–209, 229
    tail docking of, 88–89
dairy products, 1, 2–3, 175–176, 192, 200, 202, 238–239
    carbon footprint of, 84
    from grass-fed cattle, 204, 206–209, 229
    in Masai diet, 187
Danforth, Adam, 216
da Vaca, Cabeza, 186
Davis, William, 166
Dayton, Paul, 123
DDT, 85
de Araujo, Ivan, 165
death experiences in farm life, 121–122, 123
default emissions, 28–29
deforestation, 7, 10–12, 23–24, 26, 28, 52, 73
denitrification, 20
dental health, 154, 158, 184, 186
desertification, 41–48, 108–109, 113, 114
de Villalobos, Gregorio, 69
diabetes, 168, 169, 170–172, 198
    and metabolic syndrome, 156, 157
    and processed meats, 143, 144, 171
    and sugar, 153–156, 161, 162, 170
Diamond, Jared, 127, 187
diet, 129–188
    beef in. *See* beef consumption
    carbohydrates in. *See* carbohydrates
    climate impact of choices in, 56, 57
    dairy products in. *See* dairy products
    fat in. *See* fat, dietary

glycemic index in, 166, 167, 195
and heart disease, 131, 135–136. *See also* heart disease
historical, 137, 138, 183–187, 190, 193
limitations of research on, 140–142
locally produced foods in, 116, 179, 192
non-livestock substitutes in, 28–29
nutritional value in. *See* nutritional value of foods
omnivorous, 190, 192, 205, 244
optimal levels of macronutrients in, 167
processed foods in. *See* processed foods
protein in. *See* protein, dietary
sugar in. *See* sugar
and TMAO production, 172–174
vegan, 178, 195, 196, 246
vegetarian. *See* vegetarian diets
whole foods in, 191
Diet for a New America, xi
Diet for a Small Planet, 224
digestive process in ruminants, 18
Dirt (Montgomery), 25–26, 109, 243
The Dirt Cure (Shetreat-Klein), 126
Dirt to Soil (Brown), 181
DNA, 178
Doan, Jerry, 100
Dodo tribe, 232
Dole, Robert, 80, 82
domestication of cattle, 68, 127, 236
Dove, Rick, 234
draft animals, 3
dung beetles, 19, 75
Dust Bowl, 70
dyslipidemia, 155, 157

eggs, 192, 200, 204, 209, 239
Einstein, Albert, 58
elephants, 41–42
The Emissions Gap Report (2013), 22
Endangered Species Act (1994), 103
energy use, 52, 74, 83, 85, 229–230
    in alfalfa production, 215
    in food processing and transportation, 55
    life-cycle assessment of, 49
Enig, Mary, 133
environmental issues, 7–114
    biodiversity, 100–107
    climate change, 7–58

and ethics of meat eating, 229–237
overgrazing, 108–114
water use and quality, 72, 77–98
Environmental Protection Agency, 13, 16, 21, 77
erosion, 32, 41, 72–74, 108
   of cropland, 63–64, 83, 109–110, 243–244
   grasses in prevention of, 61, 62–64, 71, 98
   manure reducing, 82, 114
Escherichia coli, 205
estrogen, 150, 151
ethanol, 73
ethical issues, 224–244
European Union, 221, 222, 229
evolution, 42, 45–46, 51, 66, 113, 190, 238
exercise, 130, 187, 205

Fairlie, Simon, 8, 21, 226, 233, 242–243
Fallon, Sally, 193–194
family farms and ranches, 116–128, 225–226
farmers, 115–128
farm life, 115–128, 239–240, 245–246
fast foods, 132, 133, 135
fat, body, 153–154, 155, 166, 171–172
fat, dietary, 134–149, 167
   in high-carb diets, 168, 169
   historical, 193
   in low-carb diets, 170
   saturated. *See* saturated fats
   vegetable, 2, 131, 134–137, 179
Fat Chance (Lustig), 158, 164
faux meats, 178–182
Fearnley-Whittingstall, Hugh, 241
Feeding People is Easy (Tudge), 226, 231
feedlots, 89–91, 204, 218–220, 222, 237
fertility of soil, 21, 27, 29, 242–243
fertilizers, synthetic, 20, 37, 54, 64, 72, 83, 85
fetal bovine serum, 182
fires, 25–26, 65, 243
Fish and Wildlife Service (US), 106
fish consumption, 134, 137
flavor of meat, 205, 209–212, 213, 215
Food, Energy, and Society (Pimentel), 60, 63, 64, 81, 229, 230, 232, 243
food additives, 145, 171, 177–178
Food and Agriculture Organization, 13, 18, 30, 52–53, 54, 225
   Livestock's Long Shadow report, 8, 11, 21–24, 28–29, 52, 100

on sugar production, 160
on zinc absorption, 198
Food and Drug Administration, 135, 219, 220
food preparation, 136, 146, 174, 210
Food Rules (Pollan), 4
food security issues, 33, 38, 71
   and global hunger, 224–234
food system
   cattle in, 21, 53, 87, 93, 226–228
   consumers in, 57, 245
   cost and quality of foods in, 187–188
   energy use in. *See* energy use
   family farms in, 116
   grasslands/grasses in, 18, 37, 59, 60, 76
   locally produced foods in, 116
   pollinators in, 106–107
   processing and transportation in, 55
   production in, compared to global
      population, 225
food waste, 55, 188
food web, 238
Forages (Hughes), 71
forests, 26
   carbon in, 10, 27, 32
   and deforestation, 7, 10–12, 23–24, 26, 28, 52, 73
   fires in, 25, 65
   grazing in, 26–27
formula, infant, 164–165, 197, 201
fossil fuels, 9, 10, 17, 52, 56
   agricultural uses of, 74, 76, 85
1491 (Mann), 25
fracking process, 13
Franzluebbers, Alan, 27
freezing beef, 216
fructose, 156, 160–161, 162, 164
fungi, glomalin produced by, 34–37

Gardner, Christopher, 170
Gates, Bill, 181
genetically modified crops, 177, 179, 181
The Geography of Childhood, 123
germs, childhood exposure to, 126–127
Gillespie, David, 153
Global Change Biology, 38
global hunger, 224–234
Global Warming Potential, 14–15

glomalin, 34–37, 71, 85, 86, 110

glucose, 149, 153, 156, 161, 164, 166, 167

glutathione, 208–209

glycation, 166, 167

glycemic index, 166, 167, 195

goats, 60, 83, 109, 120, 121

Good Calories, Bad Calories (Taubes), 168

Goran, Michael, 161

gout, 154

grains, 165–168, 173, 188
    in animal feeds, 205–209, 222, 228, 229

Grandin, Temple, 219

grass buffers, 72, 75

grass-fed beef, 11–12, 31, 98–99, 204–215
    energy use in, 229–230
    nutritional value of, 54, 201, 204–209
    and soil carbon, 49, 50

grass-fed dairy cows, 83–84, 92, 204, 206–209, 229

grasslands/grasses, 59–76
    bare ground between plants in, 43–44
    biodiversity in, 63, 100–107
    carbon in, 28, 30, 36, 37–39, 50
    compared to croplands, 111
    conversion to cropland, 38, 61, 69–70, 73
    cropland converted to, 38, 39, 84
    erosion reduced in, 61, 62–64, 71, 98
    in food system, 18, 37, 59, 60, 76
    glomalin in, 36
    grazing of. *See* grazing
    historical aspects of, 17, 24–26, 42, 45–46, 51, 65–72, 104, 108–109, 111, 112–113
    livestock in management of, 243
    mowing of, 33, 61, 62
    rainfall in, 62–64, 98
    unsuitable for crop production, 230–233
    water content of soil in, 62–64, 97–98
    water use of cattle raised on, 94

grazing, 3, 7, 59–76
    beneficial effects of, 38, 40, 112, 221, 242–243, 244
    biodiversity in, 41, 43, 46, 100–107
    on cover crops, 234
    on crop residues, 230
    in crop rotations, 36, 64
    density of animals in, 39, 44, 45, 48, 113
    enteric methane emissions in, 49
    and grass-fed beef. *See* grass-fed beef

historical aspects of, 17, 24–26, 45–46, 51, 65–72, 104, 108–109, 111, 112–113
    hoof impact in, 7, 41, 44–45, 61–62, 112, 113, 114
    on land unsuitable for crop production, 230–233
    manure in, 62, 64
    and methane-oxidizing bacteria, 20
    mob grazing in, 48, 54
    mowing compared to, 33, 61, 62
    as necessary disturbance, 101, 244
    overgrazing in, 41, 45, 89, 108–114
    plant response to, 61
    on public lands, 111–112
    resting land in, 41, 42, 44, 47, 48, 89, 114
    rotational, 26
    Savory approach to, 40–48, 50–51
    in silvopasture, 26–27
    and soil carbon, 12, 31, 33, 36, 39, 48–49, 50, 51, 53, 54
    and soil fertility, 21, 242–243
    soil organisms in, 33, 62, 64
    and soil water, 48, 92

Greener Pastures (Clancy), 206, 207

greenhouse gas emissions, 8, 9
    carbon dioxide in. *See* carbon dioxide
    methane in, 9, 12–20, 28, 49, 83, 84
    nitrous oxide in, 9, 20, 83, 84
    percentage attributed to livestock, 8, 21–23, 24, 28, 29, 52

grilled meats, 146

growth hormones, 204, 205, 222

Gulf of Mexico dead zone, 83

Guns, Germs, and Steel, 127, 187

Gurven, Michael, 183, 184

gut microflora, 18, 19, 162, 172–173

Harris, Will, 49, 91, 181

Hart, Richard H., 101

Harvard School of Public Health, 168–169, 175

hay, 214–215

health issues, 129–188
    cancer, 130, 167, 174–178
    cost and quality of foods in, 187–188
    diabetes. *See* diabetes
    and germ exposure, 126–127
    heart disease. *See* heart disease
    and hunter-gatherer diet, 183–187

screen time of children in, 128
Healthy Soils Initiative, 33
healthy user bias, 140, 171, 174
heart disease, 130, 135–136, 155
    and cholesterol, 136, 139–140, 146–148, 157
    and grains, 166, 167
    and metabolic syndrome, 156, 157
    and processed meats, 143, 144
    and red meats, 173, 176, 177, 183
    and saturated fat, 131, 136, 138–140, 143, 144, 145, 157
    and sugar, 149–162
heme iron, 196, 197
herbicides, 12, 74, 83, 85, 101
herbivores, wild, 61, 101, 112–113
    function of cattle as, 101, 114, 243
    and grasslands, 24, 25, 38, 66, 101
    predators of, 25, 45, 66, 67, 68, 112–113
    reintroduction of, 67–68
    and Savory grazing method, 42, 45–46
Hereford cattle, 212
heterocyclic amines, 146
high-carb diets, 168, 169
high-density lipids (HDL), 147–148, 157
high-fructose corn syrup, 160–161
historical aspects
    of desertification, 108–109
    of foods in diet, 137, 138, 190, 193, 238
    of grazing, 17, 24–26, 45–46, 51, 65–72, 104, 108–109, 111, 112–113
    of hunter-gatherer diet, 183–187
    of livestock-borne diseases, 127
    of meat consumption, 132–133
    of sugar, 159–160, 167
holistic grazing approach, 40–48, 50–51
Holistic Resource Management (Savory), 43
honey, 159
hoof impact in grazing, 7, 41, 44–45, 61–62, 112, 113, 114
hormones, 150, 151, 204, 205, 222
Howard, Albert, 242
Howarth, Robert, 13
Hughes, H. D., 71
humus, 34, 35
hunger, global, 224–234
hunter-gatherers, 26, 183–187
hunting, 25, 26, 46, 66–67, 68
hydrogenated vegetable oils, 134–137

Hyman, Mark, 155
hypertension, 130, 131, 137, 154, 155
hyphae, 35, 36

immune system, 126–127
Impossible Foods, 180–181, 182
income in agriculture, 115, 116, 226
industrial agriculture, 4, 12, 246
    confinement operations in, 12, 16, 21–22
    ethical issues in, 235, 236–237
    manure storage in, 16
Ingalls, John James, 65
insecticides, 83, 85, 86
insects, 75, 105–107
insulin, 150, 151, 153, 155, 156, 157, 161
Intergovernmental Panel on Climate Change, 21, 31
Inuits, 185–186
Ioannidis, John, 140–142, 144
Iowa, 73, 79, 82
iron, 186, 195–197
Itzkan, Seth, 50

Jackson, Wes, 38, 81–82, 110, 111
Johnston, Bradley, 176–177
Jones, Christine, 39
Jordan, 109

Kaplan, Hillard, 183, 184
Keith, Lierre, 95–96, 97
Kennedy, Edward, 220
Kennedy, Robert F., Jr., xi, 78, 234
keratin, 29
Keys, Ancel, 131, 136, 138–139, 143, 148, 149, 152
keystone species, 75
kidney function, 150
Know Your Fats (Enig), 133
Köhler-Rollefson, Ilse, 227
Krausman, Paul, 101, 103, 107
Kremen, Claire, 107
Kresser, Chris, 140, 171–172, 202, 204
Kummerow, Fred, 136

lab-grown faux meats, 182
lactose, 164
Lakhan, Shaheen E., 162
Lal, Rattan, 38, 39, 54, 64, 74, 114

lamb consumption, 133, 134
Land Institute, 38, 81
land management, 24–26, 109–110
    animal density in, 39, 44, 45, 48, 113
    and biodiversity, 101, 107
    Savory approach, 40–48
Land Stewardship Project, 64, 93
Last Child in the Woods (Louv), 119
League for Pastoral Peoples, 227
Lee, I-Min, 130
Lee, Michael, 53–54, 97
legumes, 70, 71
leptin, 160
Let the Cowboy Ride (Starrs), 111
life-cycle assessment, 49
life expectancy, 184, 193
lifestyle
    and diabetes risk, 171, 172
    on farms/ranches, 115–128, 239–240,
    245–246
    of hunter-gatherers, 187
    sedentary, 130, 172
    and vitamin D deficiency, 168, 201–202
Lin, Steven, 184
linoleic acid, 176, 206–207, 208
liver
    in cattle, nutrients in, 194, 202, 204
    in humans, 149–150, 155, 156, 161
livestock. *See also specific animals.*
    care of. *See* animal husbandry
    and crop integration, 31, 226–227, 242
    in developing countries, 225–226
The Livestock Futures Conference, 227
Livestock's Long Shadow (FAO), 8, 11, 21–24,
    28–29, 52, 100
locally produced foods, 116, 179, 192
Louv, Richard, 119, 120, 124
low-carb diets, 168–170, 198
low-density lipids (LDL), 136, 139–140,
    146–148, 157, 166
Lundgren, Jonathan, 75
Lustig, Robert, 129, 153, 158–159, 163, 164

Malhotra, Aseem, 156–157
Mann, Charles, 25
manure, 16, 81–82, 89, 110, 114
    in confined operations, 78, 81, 82, 89, 90
    in grazing, 62, 64

and soil fertility, 21, 27, 29
Marty, Jaymee, 101, 103, 104
Masai people, 186–187
Massy, Charles, 240–241
Mayans, 193
McDougall, Christopher, 66
McGee, Harold, 146, 184, 189, 197, 205,
    213, 226
McLaughlin, Martin M., 225
Meat (Fairlie), 21, 242
meat consumption, 131
    of beef. *See* beef consumption
    ethical issues in, 224–246
    fat intake in, 131
    of processed meats, 143–144, 171, 174
    return to, 246
meat industry, 11, 217–218, 220, 221, 235, 245
meat replacement with faux meats, 178–182
megafauna, prehistoric, 66, 67
menaquinone, 200
Mencken, H. L., 58
Merck, 218–219
Merker, Moritz, 187
metabolic syndrome, 156, 157, 158, 161, 162
methane, 9, 12–20, 28, 49, 83, 84
methane-oxidizing bacteria, 19–20
microorganisms
    exposure of farm children to, 126–127
    in human gut, 162, 172–173
    in ruminant gut, 18, 19
    in soil. *See* soil organisms
milk. *See* dairy products
minerals in beef, 194
Mitloehner, Frank, 16, 28, 29
mob grazing, 48, 54
monoculture, 12, 63, 75, 116
monounsaturated fats, 144
Montgomery, David, 25–26, 32, 60, 71,
    109–110, 230, 243
moral issues, 224–244
mowing, compared to grazing, 33, 61, 62
Mozaffarian, Dariush, 147, 148, 173, 177
muscle function, 192, 194, 199, 200
mycelium, 36
mycorrhizal fungi, 35–37

Nabhan, Gary Paul, 123
National Institutes of Health, 180, 196, 198

INDEX

Native Americans, 24–26, 185, 186
natural climate solutions, 34
naturalists, 123
Natural Resources Conservation Service, 104
nature
    childhood experiences in, 118, 119, 123,
        124, 128
    disconnection from, 119–120, 123
    as model for agriculture, 241–242
Nature Conservancy, 73, 101, 103, 107, 112
nature-deficit disorder, 119
Nelson, Miriam, 200
neonicotinoids, 106
Nestle, Marion, 191, 225
New Roots for Agriculture (Jackson), 110
niacin, 194
Nichols, Kristine, 35–36, 37, 86
Niman, Bill, 118, 122, 193, 209–212, 213
Niman, Miles, 122, 125
Niman, Nicholas, 122
Niman, Nicolette Hahn, 118–120, 121–123
    Righteous Porkchop, 2, 214, 217
    as vegetarian, 192–193, 197, 235, 244
    at Waterkeeper, 78, 79, 80, 89, 91, 124, 234
Niman Farms, 215
Niman Ranch, 211, 212
nitrate, 92, 144, 146
nitrification, 20
nitrites, 171
nitrogen, 20, 33, 70–71, 215
nitrosamines, 146
nitrous oxide, 9, 20, 83, 84
nomadism, 109
nonheme iron, 196, 197
Nourishing Traditions (Fallon), 193, 201
Nourishment (Provenza), 180, 191
Nurses' Health Study, 154
nutritional value of foods, 53–54
    of beef, 53–54, 96–97, 189–216, 230
    of cheap foods, 187–188
    and global hunger issues, 228, 230
    of organ meats, 194, 203, 204
    and water use, 96–97

oak trees, 25
obesity, 2, 129, 130, 131, 137, 172
    low-carb diets in, 168
    and metabolic syndrome, 156, 157

and sugar, 153–154, 155, 158, 160, 161
    and ultra-processed foods, 180
omega-3 fatty acids, 206, 207–208
omega-6 fatty acids, 207–208
omnivores, 190, 192, 205, 244
On Food and Cooking (McGee), 184
Optaflexx, 219
organic farming, 31, 32, 114
organic matter, 27, 31, 33, 34, 35, 43, 110
organ meats, 194, 202, 203, 204
organochlorine pesticides, 85–86
Ornish diet, 169
Orr, Barron J., 10
osteopenia, 192
osteoporosis, 192, 200
ovarian cancer, 176
overgrazing, 41, 45, 89, 108–114
over-resting, 42, 47, 89
overweight, 129, 171, 172
oxidized fat, 136

pancreatic cancer, 176
pantothenic acid, 194, 203
Paris climate summit (2015), 33
People and the Earth, 61
Perlmutter, David, 167, 185
pesticides, 12, 74, 82, 85–86, 88, 97, 106
Peters, Michael, 27
phosphorus, 37, 83, 242–243
photosynthesis, 17, 37
phytonutrients, 126–127
pigs, 3, 16
    confinement of, 12, 21–22, 28, 78, 79, 80,
        82, 220
    and pork consumption, 53, 133, 134
Pimentel, David, 60, 63, 72, 81, 83, 93, 96,
    215, 229
Pimentel, Marcia, 60, 63, 81, 215, 229
plants
    diversity in grazing, 43, 46, 63, 101–102
    glutathione in, 208–209
    iron in, 196, 197
    protein in, 199, 208–209, 228
    zinc in, 198
plowing. *See* tillage
Pollan, Michael, 4, 131, 188, 236
pollinators, 106–107
pollution, 72, 77–93

polycyclic aromatic hydrocarbons, 146
polyunsaturated fats, 143, 144
pork consumption, 53, 133, 134
poultry, 16
   confinement of, 12, 21–22, 28, 78, 79, 80,
     82, 220, 239
   nutritional value of, 53
   pasture-raised, 204, 205, 209
   pounds of meat per animal, 238
   trends in consumption of, 134, 137
poverty, 225–226
prairie ecosystems, 38, 63
predators, 25, 45, 66, 67, 68, 112–113, 221
pregnancy
   in cattle, 212
   in humans, 195, 196, 197, 203
preservation of beef, 215–216
preservatives, 134–137, 144
Price, Weston, 186, 187
probiotics, 19
processed foods, 4, 12, 55, 56
   faux meats, 178–182
   meat products, 143–144, 171, 174
   sugar in, 151–152, 158–159, 163
   trans fat in, 134–137
   ultra-processed, 177–178
Project Drawdown, 27
protein, dietary, 167, 194, 198–200, 230
   in faux meats, 179
   glutathione, 208–209
   in plants, 199, 208–209, 228
Provenza, Fred, 180, 191
public lands, grazing on, 111–112
Pure, White, and Deadly, 151, 152, 157
Pure Beef (Curry), 210
Pure Farmland, 178
Purer, Edith, 102
pyridoxine, 194, 203

rainfall, 43–44, 62–64, 97, 98, 99, 109
   insufficient for crops, 231, 232
ranch life, 115–128, 239–240, 245–246
rangelands, 230–233
red meats, 1–2
   alternatives to, 28, 134
   beef. *See* beef consumption
   and cancer, 174–177
   and chronic diseases, 137, 168

cooking methods for, 146
and diabetes, 171
and heart disease, 131, 142, 143–144, 145,
   168, 173, 176, 177, 183
in hunter-gatherer diet, 183–187
Keys study on, 138
nutritional value of, 157, 190, 196–198, 202
reliability of research on, 140
and TMAO production, 172–174
trends in consumption of, 131–132, 137,
   182, 218
Reeder, Randall, 10
regenerative agriculture, 3, 4, 87, 181, 246
   beef in, 11–12
   grasses in, 64, 76
   soil biology in, 51
Reiser, Sheldon, 153
respiration, methane in, 18, 23
resting land, 41, 44, 48, 89, 114
   and over-resting, 42, 47
rewilding movement, 68
Reynolds, Kevin, 104–105
Rheaume-Bleue, Kate, 200
riboflavin, 194, 203
rice production, 16, 93, 95
rickets, 202
Rifkin, Jeremy, 108
Righteous Porkchop (Niman), 2, 214, 217
The River Cottage Meat Book (Fearnley-
   Whittingstall), 241
Roosevelt, Franklin Delano, 70
root system, 61, 62, 70
Rosen, Steven A., 108–109
rumen, 18, 19, 59
ruminants, 13, 15, 17–20, 28, 68. *See also* cattle
Running Man theory, 66–67
Ryan, Mary, 188

salt, 144, 171, 179
saturated fats, 1, 2, 131, 135, 137–149, 195
   and diabetes, 171
   in faux meats, 179
   and flavor of meat, 213
   in grass-fed and grain-fed meat, 206,
     207–208
   and heart disease, 131, 136, 138–140, 143,
     144, 145, 157
   in hunter-gatherer diet, 183–187

and keeping quality of beef, 216
in low-carb diets, 170
trends in consumption of, 133–134, 182
Savory, Allan, 40–48, 50–51, 54, 89, 102
on effective rainfall, 44, 62–63, 97
on grazing as necessary disturbance, 101
Savory Institute, 40, 100
Schaub, Bianca, 126
Schmidt, Laura, 154–155
Schwartz, Judith, 30, 39
Scottish Highland cattle, 209
Scow, Kate, 33
screen time during childhood, 127–128
seaweed supplements, 18–19
sedentary lifestyle, 130, 172
serotonin, 163
Seven Countries Study, 131, 138–139
Shapiro, Paul, 89
sheep, 3, 31, 60, 83, 109, 230
and lamb consumption, 133, 134
Shetreat-Klein, Maya, 126–127
A Short History of Farming in Britain, 90
silvopasture, 26–27
Sisson, Mark, 172
slaughter, 29, 214, 222–223, 238
age of livestock at time of, 2, 90, 205,
213, 222
fattening prior to, 90–91
fetal bovine serum harvested in, 182
soil
aggregation of, 35, 37, 61
biology of, 4, 9, 10, 34, 38, 51
as critical resource, 60
in cropland, 63–64, 83, 109–110, 243–244
erosion of. *See* erosion
fertility of, 21, 27, 29, 242–243
grasses in protection of, 62–64
organic matter in, 27, 31, 33, 34, 43, 110
Soil and Water Conservation, 62, 230, 243
Soil Association, 28, 31, 32, 34, 37, 39, 55
soil carbon, 10, 30–51
added in grazing, 12, 53, 54
in biogenic carbon cycle, 17
and carbon neutral beef, 49, 84, 181
in grasslands, 28, 30, 36, 37–39, 50, 84
in silvopasture, 27
Soil Conservation Service, 70, 71
soil organisms, 19–20, 33, 34–37, 86

biodiversity of, 81, 100
in glomalin production, 34–37, 71
in grazing, 33, 62, 64
in nitrogen fixation, 70–71
tillage affecting, 36, 38, 241
soil water content, 62–64, 97–98
in effective rain, 44, 62–63, 97
grazing strategies affecting, 48
humus affecting, 34
soil aggregation affecting, 35, 61
songbird populations, 104
soybeans, 11, 12, 24, 28, 29, 73, 83, 87, 197
spoilage, 216
Starrs, Paul, 111
Stefansson, Vilhjalmur, 185
Steinfeld, Henning, 22, 28, 227
Stewart, B. A., 38, 54, 64, 74, 114
Stokes, Michelle, 121–122
Stokes, Rob, 121
stress of cattle at slaughter time, 222–223
stroke, 130, 131, 143, 166
Strong Women, Strong Bones (Nelson), 200
subsidies, 4, 73, 75, 188
sucrose, 137, 151, 153, 156, 160, 164, 165
sugar, 2, 132, 137, 148, 149–165
and dental health, 154, 158, 184
and diabetes, 153–156, 161, 162, 170
historical aspects of, 159–160, 167
subsidized production of, 188
trends in consumption of, 159–160, 167
water usage in production of, 93–94
sugarcane, 159, 160
sun exposure, 128, 168, 201–202
supplements, nutritional, 191, 197
sustainability, 53
Sustainable Food Trust, 145
Sweden, 55, 221
Sweet Poison (Gillespie), 153

tail docking, 88–89
taste of meat, 209–212, 213, 215
Taubes, Gary, 168, 172
Taylor Grazing Act, 102, 111
Teague, Richard, 48
teeth, 154, 158, 184, 186
Teicholz, Nina, 144–145
telomere length, 178
termite methane emissions, 19

thiamin, 194, 203
Thicke, Francis, 74
Thidemann, Karl, 50
tillage, 11, 26, 36, 69–70, 111
   and biodiversity, 100, 101
   and ecosystem disruption, 241
   and erosion, 64, 69–70, 110
   grazing on land unsuitable for, 230–233
   reduced or no-till, 10, 36, 54
   and soil carbon, 10, 31
   and soil organisms, 36, 38, 241
TMAO (trimethylamine N-oxide), 172–174
trampling, 7, 41, 45, 61, 112, 113, 114
trans fats, 134–137
transportation, 9, 21, 55, 82, 222
trees, 10, 27, 32. *See also* forests
triglycerides, 147, 149, 150, 153, 156, 157, 161, 170
Trimble, Stephen, 123
trimethylamine N-oxide (TMAO), 172–174
Tudge, Colin, 226–227, 231
Tunisia, 109
Tyson Foods, 219

ultra-processed foods, 177–178, 179, 180
The Uninhabitable Earth (Wallace-Wells), 56
United Nations, 10, 22, 31
   FAO. *See* Food and Agriculture
     Organization
United States Department of Agriculture, 71, 79, 83, 112, 115, 132, 133, 138
United States Geological Survey, 85–86

Valenti, Duilio, 209
veal, 24, 82, 133, 134, 198, 205, 236
vegan diets, 12, 178, 195, 196, 246
vegetable crops, 55, 93
vegetable fats and oils, 2, 131, 134–137, 179
vegetarian diets, 192–193, 235, 244, 246
   iron in, 196, 197
   vitamin B_{12} in, 203
   zinc in, 198
The Vegetarian Myth (Keith), 95
vernal pools, 102–104, 107

vitamin A, 200
vitamin B complex, 194, 202–204
vitamin C, 194
vitamin D, 128, 167–168, 194, 200, 201–202
vitamin E, 207
vitamin K_2, 200
Volek, Jeff, 148

Wallace-Wells, David, 56
wasted food, 55, 188
water, 77–98
   in beef production, 77, 93–98, 99
   in grass-based farming, 92–93
   in grasslands, 62–64, 97–98
   quality of, 72, 77–93
   soil content. *See* soil water content
   usage of, 93–98
   in vegetable production, 93
Waterkeeper, 78, 79, 80, 89, 91, 124, 234
Waters, Alice, 211
weight gain, 192
weight loss diets, 169, 170, 171, 198
weight training, 199
wheat, 93, 94, 96, 97, 165–167
Wheat Belly (Davis), 166
White Oak Pastures farm, 49–50, 84, 181
whole foods, 191
Why We Get Fat (Taubes), 168
wildfires, 65, 243
wildlife habitat, 104, 106–107
wind erosion, 44, 62, 63, 69–70
World Bank, 56, 113
World Food Security (McLaughlin), 225
World Health Organization, 158, 195, 198, 220
Wright, Sara, 34–35

Yearbook of Agriculture (1948), 71
Yudkin, John, 137, 138, 149–153, 156, 157–158, 159, 191

Zilmax, 218–219
Zimbabwe, 41–42, 50–51
zinc, 198

ABOUT THE AUTHOR

Miles Niman

NICOLETTE HAHN NIMAN previously served as senior attorney for Waterkeeper Alliance, running its campaign to reform the industrialized production of livestock and poultry. In recent years she has gained a national reputation as an advocate for sustainable food production and improved farm animal welfare. She has been featured in *Time* magazine, *O* magazine, *Paleo* magazine, and many other publications and has appeared on the *PBS Newshour*, *The Dr. Oz Show*, and *The Diane Rhem Show*, among others. She has spoken at Yale, Stanford, UC Berkeley, and numerous other universities, and was one of just 23 speakers from around the world at the 2016 Nobel Week Dialogue about food in Stockholm, Sweden. She is the author of *Righteous Porkchop* (HarperCollins, 2009) and has written for numerous publications, including *The New York Times*, *The Wall Street Journal*, the *Los Angeles Times*, *The Huffington Post*, and *The Atlantic*. She lives on a ranch in Bolinas, California, with her husband, Bill Niman, and their two sons, Miles and Nicholas.

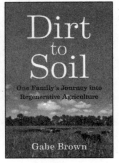
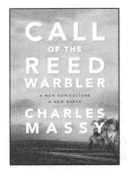